Large Language Models:

How Machines Learn Human Language

By

Susie Hala

IntelliGloss AI Education Series

Legal Disclaimer

Educational Purpose Disclaimer

"This publication is part of the IntelliGloss AI Education Series, a structured curriculum support system for AI literacy in grades 6–12."

The information presented in this publication is provided for educational and informational purposes only. This book is intended to support educators, administrators, parents, and learners in understanding the concepts, opportunities, and challenges associated with artificial intelligence in education. It is not intended to serve as legal, technical, policy, cybersecurity, or professional advice.

Readers should consult qualified professionals, institutional guidelines, and local regulations before implementing any educational programs, technologies, or policies described in this book.

Technology and Accuracy Disclaimer
Artificial intelligence is a rapidly evolving field. While every effort has been made to ensure the accuracy and reliability of the information presented at the time of publication, technologies, policies, research findings, and best practices may change over time. The author and publisher make no guarantees that all information will remain current or applicable in future technological environments.

Readers are encouraged to verify tools, platforms, and recommendations independently before using them in educational settings.

Implementation Responsibility
Schools, educators, and institutions are solely responsible for determining how and whether to implement artificial intelligence tools, strategies, or classroom activities discussed in this book. Educational environments vary widely, and policies regarding technology use, student privacy, and curriculum development differ by district, state, and country.

The author and publisher assume no responsibility for outcomes resulting from the implementation of ideas, examples, or frameworks described in this book.

Student Safety and Data Privacy
Artificial intelligence tools may involve the collection or processing of data. Educators and institutions are responsible for ensuring that any technology used complies with applicable student privacy laws, data protection regulations, and school policies, including but not limited to FERPA, COPPA, GDPR, or other relevant regulations depending on jurisdiction.

The author and publisher are not responsible for the privacy practices, security policies, or data handling procedures of third-party platforms or software mentioned in this book.

Third-Party Tools and References
This book may reference third-party technologies, organizations, research studies, or platforms for illustrative and educational purposes. Such references do not constitute endorsements or guarantees regarding the performance, reliability, safety, or suitability of those tools for educational use.

Educators and institutions should conduct their own evaluations and risk assessments before adopting any technology.

No Professional Liability
The author and publisher shall not be held liable for any direct, indirect, incidental, or consequential damages resulting from the use or misuse of the information contained in this book. This includes but is not limited to decisions related to curriculum design, technology adoption, classroom instruction, institutional policies, or educational outcomes.

Educational Interpretation
The concepts and examples provided in this book are intended to support discussion and learning. They should not be interpreted as official standards, mandatory guidelines, or universally applicable policies for artificial intelligence education.

Schools and educators are encouraged to adapt the ideas presented here according to their local policies, cultural context, educational standards, and student needs.

Author's Intent
The goal of this book is to promote responsible AI literacy, thoughtful discussion, and ethical awareness as schools prepare students for a world increasingly shaped by intelligent technologies.

The author encourages readers to approach artificial intelligence education with curiosity, caution, and a commitment to responsible learning.

Title: Large Language Model : How Machines Learn Human Language

ISBN: 978-1-972925-12-6

Published by
IntelliGloss Press
United States of America

Dedication

This book is dedicated to the next generation of learners—
the students who will grow up in a world shaped by intelligent technology,
and who deserve to understand it with clarity, confidence, and purpose.

To the educators who show up every day with dedication and heart,
guiding students through a rapidly changing world—
your role has never been more important.

To the parents and families who support learning at home,
encouraging curiosity, responsibility, and growth—
you are the foundation of every child's success.

And to my granddaughters,
whose curiosity, imagination, and future inspire me to do this work—
this book is for you.

May you not only learn how to use technology,
but understand it, question it, and shape it for the good of others.

About the Author

Susie Hala
Founder of the IntelliGloss AI Education Series

Susie Hala is an author, educator, and curriculum developer dedicated to advancing artificial intelligence literacy in education. Her work focuses on helping schools, educators, and students understand how intelligent technologies are shaping modern society—and why that understanding is essential for the next generation.

Hala is the creator of the IntelliGloss AI Education Series, a comprehensive and growing collection of educational books designed to make complex artificial intelligence concepts accessible, structured, and practical for classroom use. Her work emphasizes clarity, organization, and responsible implementation, enabling schools to introduce AI literacy without requiring advanced technical backgrounds.

In addition to her work in education, Hala brings hands-on experience working with artificial intelligence systems, including early exposure to rule-based AI technologies during her time working with AT&T. She has also developed modern digital platforms that integrate artificial intelligence, automation, and online systems, building e-commerce solutions and AI-driven applications that connect payment platforms, hosting environments, and conversational AI tools into real-world, functional systems.

This combination of foundational understanding and practical experience allows her to translate complex AI concepts into clear, accessible language for both educators and students.

Recognizing that artificial intelligence is rapidly transforming how information is created, interpreted, and used, Hala advocates for equipping students with the ability not only to use AI tools, but to understand, question, and evaluate them. Her work emphasizes critical thinking, ethical awareness, and responsible digital citizenship.

The IntelliGloss AI Education Series supports schools in teaching key areas such as artificial intelligence foundations, algorithms, machine learning concepts, and AI ethics, as well as the broader evolution of intelligent systems—from Artificial Narrow Intelligence (ANI) to Artificial General Intelligence (AGI) and Artificial Superintelligence (ASI). These resources are designed for students in grades 6–12 and for educators seeking structured, classroom-ready approaches to AI literacy.

In addition to her work in education publishing, Hala has a strong background in entrepreneurship, having built and managed multiple businesses and educational initiatives. Her experience across industries reinforces her belief that technology education must remain grounded in human values, critical thinking, and ethical responsibility.

Her mission is clear: to ensure that the next generation understands intelligent technology—not just how to use it, but how it truly works.

Preface

Two Students. Two Paths. One Future.

Before we begin, it is important to understand why learning how artificial intelligence works—not just how to use it—matters for every student.

Imagine two students who have just graduated from high school and are preparing for a job interview. The first student knows how to use artificial intelligence tools. This student can generate answers, complete assignments quickly, and rely on AI to assist with tasks. However, this student does not understand how artificial intelligence works behind the scenes.

The second student also knows how to use AI tools, but in addition, understands the foundational concepts behind them. This student has learned about data, algorithms, tokens, models, and how intelligent systems are trained and operate. This student can think critically about AI, recognize its limitations, and adapt when needed.

At first, the first student may appear faster and more efficient. However, over time, the second student demonstrates a deeper level of understanding, independence, and problem-solving ability. The second student is not only able to use artificial intelligence but also to question it, improve it, and grow with it.

This book is written for the second student.

In a world where artificial intelligence is becoming part of everyday life, students must not only learn how to use intelligent systems—they must learn how they work. True understanding leads to confidence, responsibility, and the ability to lead in an AI-driven future.

This perspective is not just theoretical—it is personal.

There was a time when I was the first student.

I knew how to use systems. I followed processes, completed tasks, and worked within structured environments that required accuracy and consistency. At the time, I did not think of these systems as artificial intelligence. They were simply part of the job—tools that had to be learned, followed, and mastered.

But looking back, I now understand something much deeper. I was working within rule-based systems—the earliest foundation of what we now recognize as artificial intelligence.

That realization changed everything.

As I began studying modern AI, including machine learning, deep learning, and intelligent systems, I started to see a clear connection between the past and the present. The structured systems I once worked in, the logic behind decision-making, and the step-by-step processes all still exist today. They have simply evolved.

Artificial intelligence did not appear overnight. It grew from rule-based systems into systems that can now learn, adapt, and assist in ways we never imagined.

I made a decision to move from simply using systems to understanding them.

That journey—from student one to student two—is the reason these books exist.

My mission is to help others bridge that same gap.

I believe that artificial intelligence should be understandable for everyone. It should not feel intimidating or out of reach. Students should feel confident exploring it. Teachers should feel prepared to introduce it. Everyday learners should have access to clear, structured explanations that make sense.

You do not need a background in technology. You do not need to know how to code. You only need the right explanation.

A Note from the Author

AI Is a Tool — Not a Decision Maker

Artificial Intelligence is becoming part of everyday life. Students use it to complete assignments, teachers use it to support instruction, and businesses use it to improve efficiency. While these tools are powerful and useful, there is an important truth that must remain clear from the very beginning:

Artificial Intelligence does not make decisions—people do.

One of the main purposes of this book is to help readers understand how AI works, what it can do, and—just as importantly—what it cannot do. AI systems are designed to process information, identify patterns, and generate responses based on data. They can assist with thinking, but they do not think independently. They can provide answers, but they do not understand consequences. They can suggest ideas, but they do not choose actions.

Every outcome that involves AI begins and ends with human involvement. A person asks the question. A person reviews the response. A person decides what to do next. Responsibility does not transfer to the machine.

In today's world, it is easy to confuse speed with intelligence and automation with authority. Because AI can respond quickly and confidently, it may appear as though it is making decisions. In reality, it is performing a task based on patterns it has learned—not exercising judgment.

This distinction matters. When people begin to rely on AI without understanding its role, they risk misunderstanding information, misusing technology, or placing trust in a system that is not capable of accountability. Clear understanding leads to responsible use.

This book was written to remove confusion and provide clarity. It is designed to help students and educators move beyond simply using AI tools and begin understanding the structure behind them. By learning how AI supports human thinking—rather than replacing it—readers can approach technology with confidence, awareness, and responsibility.

Artificial Intelligence is a powerful assistant. It can enhance learning, support creativity, and improve productivity. But it remains a tool.

The decisions, the actions, and the responsibility will always belong to the human.

— Susie Hala

Bringing Clarity to Artificial Intelligence

During a recent meeting with a school district, it became clear that one of the biggest challenges in artificial intelligence education is not the technology itself—but the way it is explained.

In that discussion, terms such as **chatbots, generative AI, agentic AI, autonomy, tools, hallucinations, and AI usage levels** were all mentioned together. While each term represents a distinct concept, they were often blended into one conversation without clear distinctions. The result was confusion, even among those who were actively trying to understand and make informed decisions.

This experience revealed an important truth:
Without clear definitions and structure, artificial intelligence becomes difficult to teach, difficult to learn, and difficult to use responsibly.

This book is designed to provide that structure.

Understanding the Core Concepts

To build a strong foundation, it is essential to separate and define the key ideas that are often confused.

Chatbot
A chatbot is a conversational interface that allows users to interact with an AI system through questions and prompts. It responds to input but does not act beyond what is directly requested.

Generative AI (Gen AI)
Generative AI refers to systems that create new content, including text, images, audio, and code. These systems generate outputs based on patterns learned from large datasets. While powerful, generative AI does not truly understand information and may occasionally produce inaccurate results.

Agentic AI
Agentic AI describes systems that can take action toward a goal. Rather than responding to a single prompt, these systems can plan, make decisions within defined limits, and carry out multi-step tasks. This introduces a higher level of functional independence.

Autonomy
Autonomy refers to how independently an AI system operates. Some systems require constant human guidance, while others can perform multiple steps with minimal input. Autonomy is not a separate type of AI, but a measure of how much control the system has in completing tasks.

Additional Terms That Shape Understanding

AI Tools
AI tools are the applications people use to interact with artificial intelligence, such as writing assistants, design platforms, and data analysis systems. These tools make AI accessible in everyday tasks.

Hallucination
A hallucination occurs when an AI system generates information that appears correct but is inaccurate or unsupported. This is a known limitation of generative systems and highlights the need for human verification.

The AI Use Continuum

Another important concept is how AI is used at different levels. Rather than viewing AI as a single capability, it is more accurate to understand it as a continuum of use:

Assistive Use
AI supports simple tasks such as answering questions, correcting grammar, or providing suggestions.

Augmented Use
AI helps expand human capability by generating ideas, organizing content, and assisting in problem-solving.

Agentic Use
AI begins to take on multi-step responsibilities, planning and completing tasks with reduced human input.

This continuum shows that AI can range from a helpful assistant to a more independent system, depending on how it is applied.

The Moment Everything Sounded Like One Conversation

During a recent discussion on artificial intelligence, I listened closely as different experts spoke about the future of AI. One referenced advanced systems beyond human intelligence, another discussed goal-driven AI systems, and the conversation shifted quickly between topics like autonomy, tools, and system behavior. At first, it sounded like a single, unified discussion. But the more I listened, the more I realized something important.

They were not all talking about the same thing.

One part of the conversation focused on future possibilities—high-level systems that do not yet exist. Another part focused on current technologies that are already being used today. At the same time, different terms were being used to describe how AI behaves, how it makes decisions, and how much independence it may have. These ideas were layered together in a way that made them sound connected, even when they were not.

This experience revealed a major challenge in understanding artificial intelligence today. Many discussions combine different types of AI, different levels of intelligence, and different behaviors into one conversation. Without clear structure, these ideas begin to blur together. For someone trying to learn, this can feel overwhelming and confusing.

This book is built to solve that problem.

Artificial intelligence is not one single concept. It is a system made up of multiple layers. Some layers describe how intelligent a system is. Other layers describe how the system behaves. Still others describe how much control or independence the system has. When these layers are separated and explained clearly, AI becomes much easier to understand.

The goal of this book is to bring clarity where confusion often exists. Instead of mixing everything into one explanation, each concept is presented in a structured and organized way. This allows students, educators, and readers to see how the pieces fit together without losing sight of what each part actually means.

Understanding artificial intelligence begins with understanding its structure.

Once that structure is clear, the conversation becomes clearer as well.

Why Clarity Matters

When these concepts are not clearly defined, they merge into what can be described as a "conceptual blur." This can lead to misunderstandings about what AI can and cannot do, as well as unrealistic expectations or misuse in educational settings.

Clear understanding helps ensure that:

Students remain active learners, not passive users

Educators maintain control over instructional outcomes

AI is used as a support tool rather than a replacement for thinking

The Purpose of This Book

This book is built on a simple but essential principle:

Students should not only use artificial intelligence—they should understand it.

By clearly defining core concepts such as chatbots, generative AI, agentic AI, autonomy, and the continuum of AI use, this curriculum provides a structured pathway for developing both technical understanding and critical thinking.

Artificial intelligence is not a single tool or system. It is a collection of capabilities that must be understood in context.

Clarity is the foundation of responsible innovation.

Structuring Artificial Intelligence to Create Clarity and Understanding

Why Structure Is Necessary

Artificial intelligence is often presented as one unified concept. In many discussions, different ideas such as intelligence level, system behavior, and autonomy are combined into a single explanation. This can make AI appear more complex and confusing than it actually is.

Clarity begins when these ideas are separated and organized.

Structuring artificial intelligence allows each concept to be understood on its own before being connected to the larger system.

The Problem Without Structure

When AI is not structured, conversations may include multiple terms that sound related but describe entirely different things. For example, a discussion may move quickly between advanced future systems, current tools, and system behaviors without clearly distinguishing between them.

This creates the impression that all forms of AI operate at the same level and in the same way, which is not accurate.

Without structure, understanding becomes difficult.

The Solution: A Structured Approach

Artificial intelligence becomes clearer when it is organized into distinct categories.

Level explains how intelligent a system is.
Behavior explains how the system operates.
Control explains how much independence the system has.

Each category answers a different question. Together, they form a complete picture of how AI systems function.

Clarity Through Separation

A system can operate at a current level of intelligence while demonstrating different behaviors or levels of independence. For example, a system that generates responses and a system that takes steps toward completing a task may operate at the same intelligence level but differ in how they behave.

This distinction becomes clear only when structure is applied.

Instructional Insight

When teaching artificial intelligence, it is important to avoid presenting all concepts at once. Separating ideas into categories allows learners to focus on one dimension at a time, building understanding step by step.

Structured learning leads to deeper comprehension.

Key Takeaway

Artificial intelligence is not one single idea. It is a system made up of multiple parts. Structuring those parts creates clarity, improves understanding, and allows learners to see how the system truly works.

Reflection Question

How does separating intelligence level, behavior, and control make artificial intelligence easier to understand?

The IntelliGloss Instructional Approach

The IntelliGloss AI Education Series is designed to help students move beyond simply using AI tools and begin understanding artificial intelligence as a complete system.

Students learn how AI is built, how it operates, and how it connects to the real world.

A Message to Students and Readers

If you have ever felt that technology was too complicated…
If you have ever been confused by systems that were never clearly explained…

I want you to know:

You are capable of understanding artificial intelligence.

Sometimes, the challenge is not the concept itself—it is how it has been presented.

This series was created to change that.

A Message to Educators

To the educators using this material:

These books are designed to support you in introducing artificial intelligence in a way that is clear, structured, and meaningful for students.

AI is becoming part of every industry, and students deserve more than just exposure to tools—they deserve understanding.

This series aims to provide that foundation.

Final Reflection

Every experience we have—especially the challenging ones—prepares us for something greater.

The systems I once worked in taught me discipline, structure, and attention to detail. Today, those same lessons allow me to break down complex ideas and make them understandable for others.

That is what this series represents:

Turning complexity into clarity.

Why Artificial Intelligence Education Matters for Every Student

Why AI Literacy Is Essential for Every Student

Artificial intelligence is no longer a future concept or a specialized technology used only by experts. It is already part of everyday life for students of all ages. From smartphones and search engines to navigation apps, social media, streaming platforms, and educational tools, AI systems influence how information is accessed, decisions are made, and learning takes place.

Because AI is embedded in daily life, understanding it is no longer optional. It is a foundational skill, similar to digital literacy and critical thinking. Students do not need to become programmers or engineers, but they do need to understand what AI is, how it works at a basic level, and how it affects their lives.

AI education should not be limited to computer science classrooms. Every student interacts with AI systems daily. Teaching students how these systems function empowers them to become informed users rather than passive consumers of technology.

This book is designed to make artificial intelligence understandable, approachable, and relevant. It focuses on clarity, responsibility, and real-world understanding to help students prepare for a future where humans and intelligent systems work side by side.

IntelliGloss AI Education Series

Teaching how AI actually works—from hardware to power to global systems.

Why AI Education Matters

Artificial intelligence is already shaping how students learn, write, research, and solve problems.

Yet many students—and educators—are using AI tools without understanding how these systems actually work.

Schools need more than access to technology.
They need structured, responsible AI education.

Our Approach

The IntelliGloss AI Education Series is designed to help students move beyond simply using AI tools and begin understanding artificial intelligence as a complete system.

Students learn how AI is built, how it operates, and how it connects to the real world.

The IntelliGloss Framework

Our books introduce AI through a clear and connected system:

Chip Family — the hardware that powers AI

Watts Family — the rate at which AI uses power

Energy Family — the total power AI consumes over time

Infrastructure Family — the global systems that support AI

This framework helps students understand AI from the inside out.

Built for the Classroom

Each book includes:

Structured chapters aligned with semester learning

Clear explanations designed for student understanding

Reflection sections with learning objectives

Worksheets, quizzes, and answer keys

Teacher Implementation Guides

Supporting Schools and Educators

The IntelliGloss series supports:

Responsible and ethical AI use

Student data awareness and safety

Critical thinking and digital literacy

Classroom-ready instruction for grades 6–12 and beyond

What This Book Is—and What It Is Not

Many people today are becoming familiar with artificial intelligence through tools like ChatGPT, Microsoft Copilot, Canva, and other AI-powered applications. These tools can help with writing, answering questions, creating images, and completing everyday tasks.

In addition to these tools, you may also hear terms such as **AI agents, agentic AI, autonomous systems, and generative AI**. While these may sound complex, they are all part of the same evolving landscape of artificial intelligence technologies.

This book is not a guide on how to use those tools.

Instead, it explains what makes those tools possible.

Behind every AI tool is a system made up of multiple layers working together. At the most basic level, information is represented as bits and processed through powerful computer chips. That information is organized into data, broken into tokens, and then processed by algorithms and models that allow the system to recognize patterns, generate responses, and perform tasks.

These components—chips, bits, tokens, data, algorithms, and models—work together as a complete system. While users may only see the tool, the real intelligence comes from how these underlying parts interact behind the scenes.

Artificial intelligence can also be understood in terms of different levels of capability. Most of the AI systems used today fall under **Artificial Narrow Intelligence (ANI)**, which is designed to perform specific tasks, such as answering questions or generating content. More advanced forms, such as **Artificial General Intelligence (AGI)** and **Artificial Superintelligence (ASI)** represent future possibilities where AI could match or exceed human-level thinking. While these advanced levels are still theoretical, understanding the differences helps provide a broader view of where AI is today and where it may be heading.

For students, parents, and educators, this distinction is important. Learning how to use AI tools is helpful, but understanding how they work—and where they fit within the larger AI landscape—builds stronger thinking, better decision-making, and more responsible use.

Most importantly, artificial intelligence does not make decisions on its own. It does not take responsibility for actions or outcomes. AI systems generate responses based on data and patterns, but it is always the human who decides what to do with that information.

This book is designed to make these underlying systems clear and accessible, even for readers who are new to artificial intelligence.

Artificial intelligence is still in its early stages of development. As these technologies continue to grow, those who understand the foundation behind them will be better prepared to adapt, question, and use them wisely.

Who This Book Is For

This book is designed for educators, school leaders, and curriculum developers guiding students in an age of rapidly advancing technology.

Artificial intelligence is already influencing how students learn, research, and communicate. Yet many educators have not received formal training in how these systems work or how they should be addressed in the classroom.

This book helps bridge that gap by providing clear explanations and practical frameworks for understanding artificial intelligence in an educational setting.

How Schools Can Use This Book

This book supports schools in developing a structured approach to artificial intelligence literacy.

It can be used for:

Professional development for educators

Classroom discussions and activities

Curriculum support across subject areas

Policy and academic integrity conversations

Rather than focusing on specific tools, this book emphasizes lasting principles that help students understand, evaluate, and use AI responsibly.

Closing Perspective

Artificial intelligence will continue to evolve. With the right knowledge and leadership, schools can ensure that students not only use these technologies, but understand them, question them, and apply them responsibly.

IntelliGloss is more than a book series. It is a structured approach to preparing students for a future shaped by artificial intelligence.

Rather than focusing only on how to use AI tools, this series helps students understand how AI actually works—from its foundational building blocks to real-world applications.

"Before exploring the technical structure of artificial intelligence, it is equally important to understand the human principles that guide its use."

Human Values Family

Definition

The Human Values Family in artificial intelligence represents the guiding principles that shape how AI systems are created, used, and integrated into society. These values include ethics, responsibility, trust, and human well-being.

Artificial intelligence may be built on data, algorithms, and computational systems, but its true impact is measured by how it affects people. For this reason, AI is not only a technical system—it is a human-centered system.

Key Concepts

The Human Values Family exists to ensure that artificial intelligence serves humanity in a positive and responsible way. While AI systems can process information and generate outcomes, they do not possess judgment, empathy, or moral understanding.

This means that human values must guide every stage of AI development—from design to deployment.

AI reflects the intentions, decisions, and assumptions of the people who build and use it. If those values are not carefully considered, AI systems can unintentionally produce biased, harmful, or misleading results.

Understanding human values in AI is not optional—it is essential.

Core Elements of the Human Values Family

Ethics
Ethics in AI refers to making decisions that are fair, just, and respectful of human rights. This includes minimizing bias, protecting privacy, and ensuring transparency in how systems operate.

Responsibility
Responsibility means that humans remain accountable for AI systems. Developers, educators, organizations, and users must take ownership of how AI is used and the outcomes it produces.

Trust
Trust is critical for the successful adoption of AI. People must feel confident that AI systems are reliable, safe, and aligned with their best interests.

Human Well-Being
AI should enhance human life—not replace, harm, or diminish it. This includes supporting learning, improving access to information, and creating opportunities while maintaining human dignity.

Real-World Connection

Artificial intelligence is already influencing how people learn, communicate, and make decisions. From search engines to classroom tools, AI is shaping daily experiences.

Without clear human values guiding these systems, there is a risk of reinforcing misinformation, bias, or unrealistic expectations.

For students, this means learning not only how to use AI tools, but how to question them, evaluate their outputs, and understand their limitations.

Applications in Education

In the classroom, the Human Values Family helps students develop critical thinking and ethical awareness. Teachers can use this framework to guide discussions about fairness, responsibility, and the impact of technology on society.

Students can explore questions such as:

How should AI be used responsibly?

What makes an AI system fair?

How does technology influence human behavior?

This approach supports both technical understanding and responsible decision-making.

Benefits

Teaching human values in AI helps students:

Develop ethical awareness

Strengthen critical thinking

Understand the societal impact of technology

Become responsible users and future creators of AI

It ensures that learning about AI goes beyond functionality and includes responsibility.

Challenges

Human values are not always universal. Different cultures, communities, and individuals may have different perspectives on what is considered fair or ethical.

Additionally, AI technology is evolving rapidly, often faster than policies and guidelines can keep up. This makes it essential for educators and students to stay informed and adaptable.

Summary

The Human Values Family reminds us that artificial intelligence is not just about machines—it is about people.

By focusing on ethics, responsibility, trust, and human well-being, students can learn to use AI thoughtfully and responsibly.

Understanding these principles helps prepare learners to navigate a world where artificial intelligence plays an increasingly important role in everyday life.

Executive Summary

IntelliGloss AI Education Series

Teaching How Artificial Intelligence Actually Works

Artificial intelligence is already shaping how students learn, write, research, and make decisions. Yet in many classrooms, students are being introduced to AI tools without understanding how these systems actually function.

This creates a critical gap.

Students may become efficient users of technology, but without foundational knowledge, they are not equipped to think critically, evaluate outputs, or use AI responsibly. As AI continues to expand across every industry, this gap will only grow more significant.

The **IntelliGloss AI Education Series** was created to address this challenge.

The Problem

Most current approaches to AI in education focus on tool usage rather than system understanding. As a result:

Students rely on AI without understanding its limitations

Educators lack structured frameworks for teaching AI concepts

Key terms such as generative AI, agentic AI, and autonomy are often introduced without clear definitions

Responsibility and ethical use are not consistently emphasized

Without clarity, artificial intelligence becomes difficult to teach, difficult to learn, and difficult to manage in a school environment.

The Solution

The IntelliGloss AI Education Series provides a structured, classroom-ready approach to artificial intelligence literacy for grades 6–12.

Rather than focusing on individual tools, this series teaches students how AI works as a complete system—from foundational components to real-world applications.

Students learn to move beyond surface-level interaction and develop true understanding.

The IntelliGloss Framework

The series introduces AI through a clear and connected structure, including:

- Foundational building blocks such as data, algorithms, tokens, and models
- System-level understanding through categories such as intelligence level, behavior, and control
- Real-world context through applications, limitations, and responsible use
- Human-centered learning through ethics, responsibility, trust, and student safety

This structured approach transforms artificial intelligence from a confusing concept into an understandable system.

Instructional Design

Each book in the series is designed for practical classroom use and includes:

- Structured chapters aligned with instructional pacing
- Clear, student-friendly explanations
- Reflection sections to reinforce understanding
- Worksheets, quizzes, and answer keys
- Teacher implementation support

No prior technical background is required for educators or students.

Guiding Principle

A central message of the IntelliGloss series is:

Artificial Intelligence is a tool—not a decision maker.

Students learn that while AI can support thinking and provide information, all decisions, actions, and responsibilities remain with humans. This principle reinforces critical thinking, accountability, and responsible technology use.

Why It Matters

AI literacy is no longer optional. It is a foundational skill.

Students who understand how AI works will be better prepared to:

Think critically about information

Adapt to emerging technologies

Use AI responsibly and ethically

Succeed in an increasingly AI-driven world

Our Goal

The IntelliGloss AI Education Series is designed to help schools move beyond simply using artificial intelligence tools and toward a deeper understanding of how these systems actually work.

By providing clear structure, defined concepts, and real-world context, this series equips students and educators with the knowledge needed to think critically, evaluate information, and use AI responsibly.

The goal is not just to introduce technology—but to build understanding, confidence, and accountability in an AI-driven world.

This is not just about learning to use AI—it is about preparing students to understand it, question it, and lead with it.

Clarity leads to confidence.
Understanding leads to responsible use.

Mission Statement

The mission of the IntelliGloss AI Education Series is to make artificial intelligence clear, structured, and accessible for every student and educator.

In a world where AI is rapidly transforming how information is created, used, and understood, this series is designed to move beyond surface-level interaction and provide a deeper understanding of how intelligent systems actually work.

Through clear explanations, organized frameworks, and real-world context, IntelliGloss equips learners with the ability to think critically, evaluate information, and engage with technology responsibly.

This mission is grounded in a simple belief:
students should not only learn how to use artificial intelligence—they should understand it.

By building both technical awareness and human-centered understanding, the IntelliGloss AI Education Series prepares students to navigate, question, and shape a future increasingly influenced by intelligent technologies.

The goal is not just to keep up with artificial intelligence—but to understand it, guide it, and use it responsibly.

Vision Statement

The vision of the IntelliGloss AI Education Series is to create a future where artificial intelligence is clearly understood, responsibly used, and thoughtfully integrated into education and everyday life.

In this future, students are not passive users of technology, but informed thinkers who understand how intelligent systems work, recognize their limitations, and apply them with purpose and responsibility.

Educators are equipped with clear frameworks and structured resources that make artificial intelligence accessible, teachable, and meaningful across all subjects—not just in technical fields.

Schools become environments where technology is not simply adopted, but understood—where students are encouraged to question, evaluate, and engage with AI in ways that support learning, creativity, and ethical awareness.

As artificial intelligence continues to evolve, this vision supports a generation that is prepared not only to adapt to change, but to lead it with knowledge, confidence, and integrity.

The future of artificial intelligence will not be defined by technology alone—but by the understanding and responsibility of those who use it.

Connecting AI Understanding to Cybersecurity Awareness

Definition

Understanding artificial intelligence is the first step toward understanding cybersecurity. While cybersecurity focuses on protecting systems, data, and users, artificial intelligence explains how those systems are built, how they function, and where they may be vulnerable.

The Connection

Every AI system is made up of components such as data, algorithms, models, and infrastructure. These same components are also the targets of cybersecurity threats.

When students understand:

How data is collected and stored

How algorithms make decisions

How AI models process information

How systems operate across frontend and backend layers

They begin to recognize where risks can exist and how systems can be misused.

Why This Matters

In today's digital world, students are not just users of technology—they are participants in complex systems powered by artificial intelligence.

Without understanding how these systems work:

Technology becomes something they trust without question

Risks become harder to recognize

Decisions are made without awareness of consequences

With understanding:

Students become more aware of how their data is used

They recognize how systems can be manipulated

They make more informed and responsible choices

Building Digital Readiness

Digital readiness is not only about using tools. It is about understanding the systems behind those tools.

By learning how AI works, students develop:

Awareness of digital environments

Confidence in navigating technology

A foundation for future learning, including cybersecurity

Conclusion

Cybersecurity begins with understanding.

Before students can protect systems, they must first understand how those systems are built and how they operate. This is why learning artificial intelligence is not only about innovation—it is also about responsibility.

The Foundation of Artificial Intelligence

Definition

The foundation of Artificial Intelligence (AI) refers to the core building blocks that allow AI systems to function. These building blocks work together to help machines learn from information, recognize patterns, and make decisions. AI is not magic—it is a system built on structured components that operate behind the scenes.

Understanding the Foundation in Simple Terms

Artificial Intelligence can be compared to a house. Before a house is complete, it must be built on a strong foundation. This foundation includes concrete, wiring, plumbing, and support structures that are not always visible but are essential for the house to function properly.

In the same way, AI systems rely on foundational components that are not visible to users but are necessary for AI tools to work. When people use AI applications such as chatbots or design tools, they are interacting with the surface. The true power of AI comes from what is underneath.

Key Components of the AI Foundation

1. Data (The Information)

Data is the starting point of all AI systems. It includes text, images, numbers, audio, and other forms of information. AI systems learn by analyzing patterns within this data.

2. Algorithms (The Instructions)

Algorithms are step-by-step instructions that tell the AI system how to process data. They guide how the system learns, identifies patterns, and produces results.

3. Models (The Trained System)

A model is created when data and algorithms are combined through a training process. The model acts as the "brain" of the AI system, allowing it to make predictions and generate responses.

4. Computing Power (The Engine)

AI systems require powerful computers to process large amounts of data. This includes specialized hardware such as processors and graphics processing units (GPUs) that allow AI to function efficiently.

5. Infrastructure (The Environment)

Infrastructure refers to the systems that support AI operations, including servers, storage, networks, and energy systems. Data centers are a key part of this infrastructure, providing the physical space where AI systems operate.

How These Components Work Together

AI systems function by combining all foundational components into a single process. Data is collected and analyzed using algorithms. This process creates a model that can recognize patterns and make decisions. The entire system is powered by computing machines and supported by infrastructure.

Why Understanding the Foundation Matters

Many people use AI tools without understanding how they work. This can lead to confusion, misuse, or over-reliance on technology. By understanding the foundation of AI, students and educators can:

- Develop critical thinking skills
- Use AI tools more responsibly
- Recognize the limitations of AI systems

Build a deeper understanding of modern technology

Real-World Example

When a student asks a chatbot a question, the response does not come from "thinking" like a human. Instead, the AI system uses a trained model that has learned patterns from large amounts of data. It follows algorithms to predict the most appropriate answer, using powerful computers and infrastructure to deliver the response quickly.

This process is also influenced by how the question is asked. The words, structure, and clarity of a question—often referred to as a *prompt*—can affect the quality of the response. Learning how to communicate effectively with AI systems is an important skill known as prompt engineering. For further understanding, students and educators may explore *What to Say When You Talk to Yourself: A Guide to Prompt Engineering* by Susie Hala, which explains how carefully structured prompts can guide AI responses more effectively.

Conclusion

The foundation of Artificial Intelligence is built on data, algorithms, models, computing power, and infrastructure. These components work together to create systems that can learn, predict, and assist humans. Understanding this foundation helps students move beyond simply using AI tools and toward truly understanding how artificial intelligence works.

The Foundational Families of Artificial Intelligence

Introduction

Artificial intelligence is often described as a powerful technology, but in reality, it is not a single system. It is a combination of multiple components working together to create intelligent behavior.

To better understand how artificial intelligence works, it is helpful to break it down into organized groups. In the IntelliGloss framework, these groups are called **Foundational Families of Artificial Intelligence**.

Each family represents a key part of how AI systems are built and operate. Some families focus on how information is created and stored. Others focus on how machines learn, make decisions, process language, or perform tasks. Additional families represent the physical hardware, energy systems, and global infrastructure that make artificial intelligence possible.

When these families work together, they form a complete AI system.

This approach allows students and educators to move beyond simply using artificial intelligence tools and begin to understand what is happening behind the scenes. Instead of viewing AI as a mystery, learners can see it as a structured system made up of clearly defined components.

The diagram on the next page provides a visual overview of these families and how they connect to form the foundation of artificial intelligence.

By learning each family step by step, students develop a deeper understanding of how AI systems process information, learn from data, and interact with the world.

Understanding the Foundational Families of Artificial Intelligence is the first step toward true AI literacy.

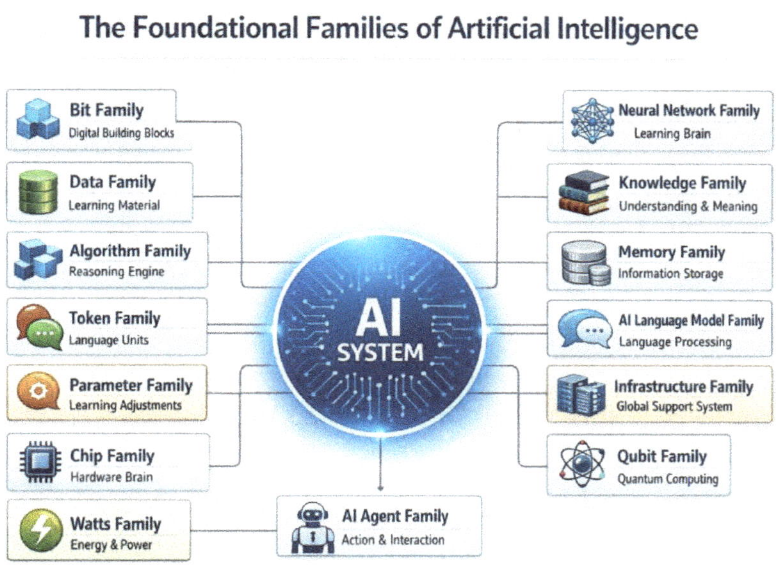

The Foundational Families of Artificial Intelligence

Family	Role	Description
Bit Family	Digital Building Blocks	The smallest unit of digital information (0 or 1). All computing systems begin with bits.
Algorithm Family	Reasoning Engine	Step-by-step instructions that tell computers how to solve problems and process information.
Token Family	Language Units	Pieces of text used by AI models to understand and generate human language.

Family	Role	Description
Parameter Family	Learning Adjustment System	Adjustable numerical values inside AI models that allow them to learn patterns from data.
Data Family	Learning Material	The information AI studies in order to learn patterns, relationships, and knowledge.
Logic Family	Decision Rules	The reasoning structures that allow AI systems to make decisions and evaluate conditions.
Memory Family	Information Storage	Systems that store and retrieve information so AI models can access past knowledge and context.
Knowledge Family	Understanding & Meaning	Structured information that allows AI to represent facts, relationships, and concepts.
Neural Network Family	Learning Brain	Interconnected layers of artificial neurons that enable AI systems to recognize patterns, learn from data, and make predictions.
AI Model Family	Intelligence Engine	The trained system that uses data, parameters, and neural networks to perform tasks such as prediction, classification, and generation.
AI Translator Architecture Family	Language Conversion System	Systems that transform input into output, enabling AI to convert text, speech, or images into meaningful responses.
AI Agent Family	Action and Task Execution System	AI systems designed to perceive information, make decisions, and carry out tasks autonomously or semi-autonomously.
Chip Family	Hardware Brain	Physical processors (GPUs, TPUs, CPUs, NPUs) that perform AI computations.
Watts Family	Energy Power	The electrical power required to run AI hardware and data centers.
AI Energy Family	Energy Source System	The total energy consumed over time to operate AI systems, including training and real-time usage.
Infrastructure Family	Global Support System	Data centers, internet networks, cloud systems, and global infrastructure that allow AI to operate and scale.
Ingredient Family	System Composition	The essential components that come together to create an AI system, including data, algorithms, models, hardware, and energy.
Qubit Family	Quantum Information Units	Quantum bits that can represent multiple states simultaneously, forming the foundation of quantum computing.

The Bit Family

Introduction

The Bit Family represents the starting point of all digital technology. Every computer system, application, and artificial intelligence model is built from bits—the smallest unit of information in computing.

A bit can hold one of two values: 0 or 1. While this may seem simple, these two values form the foundation of everything digital. When bits are combined and organized, they create larger units of data that allow computers to store text, display images, play videos, and run complex systems.

From a single bit to massive data systems, all digital information follows this same structure. This is how computers are able to represent and process the world in a form they can understand.

In artificial intelligence, bits are essential because they store the data that models learn from, process the calculations that drive decision-making, and support the systems that generate responses.

Understanding the Bit Family helps students see how simple binary signals grow into powerful technologies. It reveals that behind every advanced AI system is a foundation built from the most basic building blocks of information.

Bit Family — Visual Understanding of Data Size

How Much Data Can Each Unit Hold? (Simple Examples)

Unit	Size	Photos (Approx.)	Video (Approx. Hours)
Bit	1 bit	N/A	N/A
Byte	8 bits	1 character	N/A
Kilobyte (KB)	~1,000 bytes	Part of a paragraph	N/A
Megabyte (MB)	~1,000 KB	~1 photo	~1 minute of video
Gigabyte (GB)	~1,000 MB	~250 photos	~1–2 hours of video
Terabyte (TB)	~1,000 GB	~250,000 photos	~1,000–2,000 hours of video
Petabyte (PB)	~1,000 TB	~250 million photos	~1–2 million hours of video
Exabyte (EB)	~1,000 PB	~250 billion photos	~1–2 billion hours of video
Zettabyte (ZB)	~1,000 EB	~250 trillion photos	~1–2 trillion hours of video
Yottabyte (YB)	~1,000 ZB	~250 quadrillion photos	~1–2 quadrillion hours of video

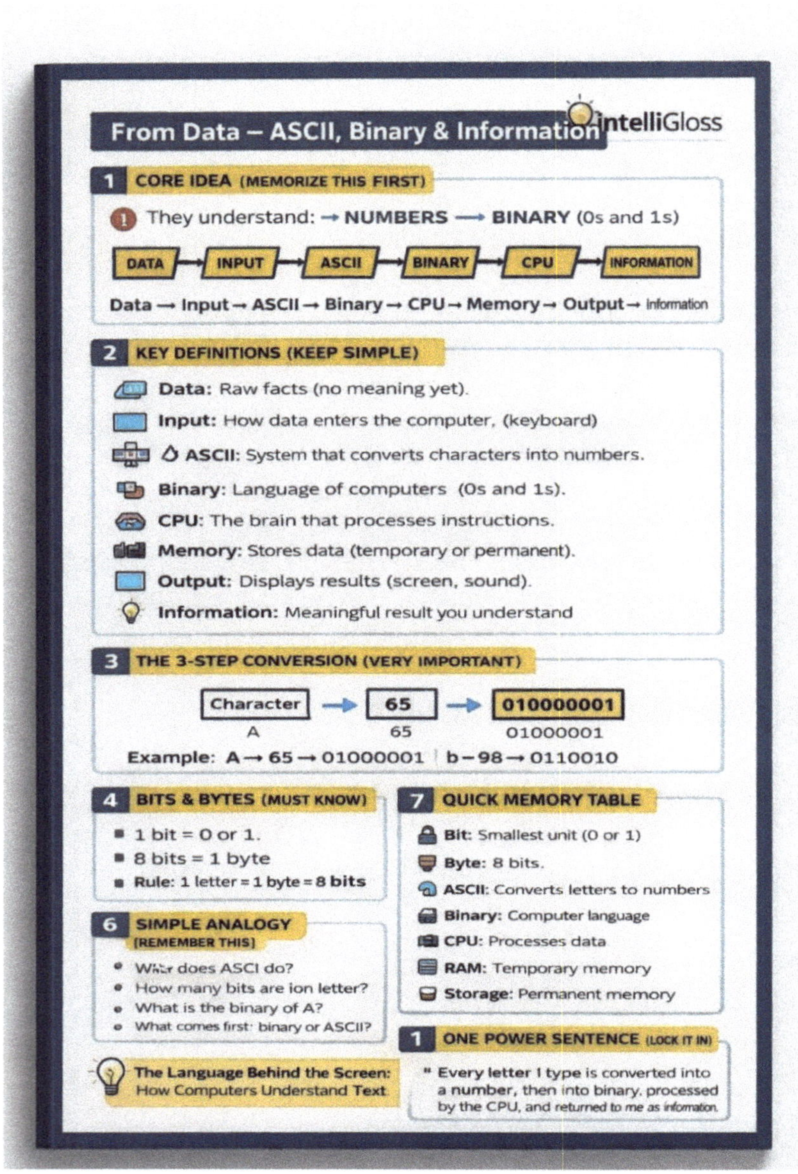

The Algorithm Family

Definition and Overview

The **Algorithm Family** represents the *thinking engine* of artificial intelligence — the set of mathematical instructions that guide how AI analyzes data, learns patterns, and makes decisions.

An **algorithm** is a sequence of steps designed to solve a problem or complete a task. In AI, algorithms tell the system *how to learn*, *how to predict*, and *how to improve* with every iteration.

In simple terms:

The Algorithm Family is the *logic of intelligence* — the invisible code that transforms data into decisions and predictions into progress.

🧠 How Algorithms Power AI

Every AI system — from simple chatbots to large-scale deep learning models — relies on algorithms as its foundation.
They determine how the model:

- Processes and cleans data,
- Identifies relationships between variables,
- Learns from feedback, and
- Improves over time.

Without algorithms, even the most powerful dataset or neural network would remain static and lifeless.

In essence, **algorithms are the teachers** that train AI how to think.

⚙️ *Why the Algorithm Family Matters*

Algorithms are what separate automation from intelligence.
They allow machines to **reason**, **adapt**, and **learn** — not just follow commands.
For prompt engineers, understanding algorithmic behavior helps explain *why* AI sometimes misunderstands a question, repeats patterns, or changes tone after feedback.

Every AI response you receive follows an underlying process — a pattern recognition journey from input to output.

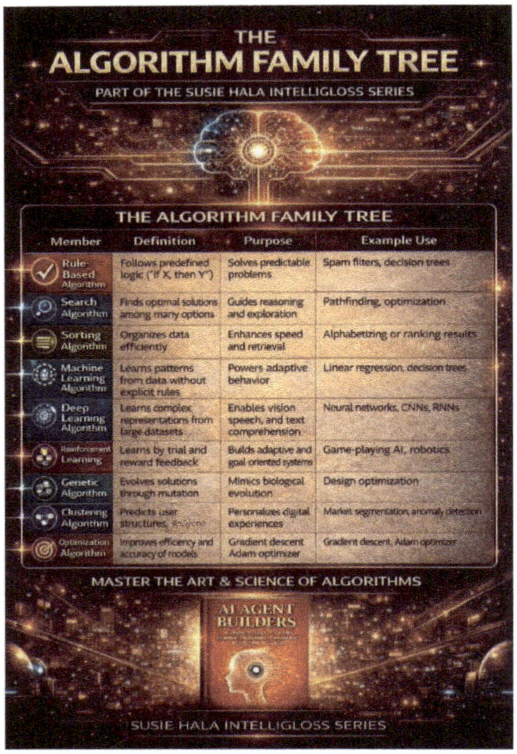

🌍 Analogy for Everyday Readers

Imagine AI as a chef in a kitchen:

The **Dataset Family** provides the ingredients.

The **Knowledge Family** gives the recipes.

The **Algorithm Family** is the *cooking process* — mixing, heating, tasting, and adjusting until the dish is perfect.

Just as cooking methods determine the quality of food, algorithms determine the quality of intelligence.

⚡ Algorithm Efficiency and Energy Use

Not all algorithms are equal.
Some are **fast but shallow**, while others are **deep but energy-intensive**.
Training large AI models requires algorithms that can handle billions of calculations per second — powered by the **Watts Family** (energy) and executed through the **Chip Family** (hardware).

The design of efficient algorithms directly affects sustainability in AI — minimizing power consumption while maximizing accuracy and speed.

🧩 Algorithms and Prompt Engineering

For prompt engineers, algorithms explain *how* the AI arrives at its answers.
When a model gives inconsistent or off-topic results, it's often due to:

Misinterpretation of input signals (input encoding),

Bias in training algorithms, or

Overfitting to certain data patterns.

Knowing this helps you craft prompts that guide the AI's decision-making — not just its output.

💬 **Quick Takeaway: The Algorithm Family is the reasoning heart of AI.**
It transforms information into understanding and enables machines to think, learn, and evolve like digital problem-solvers.

Absolutely, partner — this is a powerful section in your book because **tokens are where AI meets language**. I'll keep this aligned with your style: clear, structured, and school-district ready.

Token Family Introduction

The Token Family represents one of the most important bridges between human language and artificial intelligence. While humans communicate using words, sentences, and meaning, AI systems do not understand language in the same way. Instead, they rely on tokens—small units of text that allow machines to process, analyze, and generate language step by step.

A token can be a word, part of a word, a character, or even punctuation. When a student types a sentence into an AI system, that sentence is not read as a whole idea. It is first broken down into tokens. These tokens become the input that the AI model uses to recognize patterns, predict outcomes, and generate responses.

The Token Family works closely with several other foundational families in artificial intelligence. It connects directly with the Data Family, which provides the text used for training, and the Neural Network Family, which processes token patterns to produce intelligent outputs. It also plays a key role in the AI Translator Architecture Family, where tokens are transformed, analyzed, and reassembled into meaningful responses.

Understanding tokens helps students see what is happening behind the scenes when they interact with AI tools. Instead of viewing AI as a "magic system," they begin to understand that every response is built from sequences of tokens processed through mathematical models.

In simple terms, if language is what humans speak, tokens are what AI understands.

Token Family Breakdown Chart

Component	Role	Description	Example
Tokenization	Input Conversion	The process of breaking text into smaller units (tokens) that AI can process	"Artificial Intelligence" → "Artificial" + "Intelligence"
Tokens	Language Units	The individual pieces of text used by AI models for understanding and generation	"AI", "learn", "ing", "."
Subword Tokens	Efficiency Units	Words split into smaller parts to handle unknown or complex vocabulary	"unbelievable" → "un", "believ", "able"
Character Tokens	Fine-Grained Units	Text broken down into individual characters for detailed processing	"AI" → "A" + "I"
Vocabulary (Token Set)	Token Library	The complete set of tokens an AI model recognizes and uses	GPT models have thousands to millions of tokens
Encoding	Token Mapping	Converting tokens into numerical representations for computation	"AI" → [1234, 5678] (example IDs)
Decoding	Output Reconstruction	Converting numerical outputs back into readable text	[1234, 5678] → "AI system"
Context Window	Memory Limit	The number of tokens an AI can process at one time	Example: 8,000 tokens ≈ several pages of text
Token Embeddings	Meaning Representation	Mathematical vectors that capture relationships between tokens	"king" and "queen" have similar embeddings
Token Prediction	Language Generation	The process of predicting the next token in a sequence	"AI is" → predicts "powerful"

The Token Family reminds us that behind every intelligent response is a structured sequence of tokens working together. What feels like natural conversation to humans is, for AI, a carefully processed stream of language units transformed into meaning.

The AI Ecosystem Family

The AI Ecosystem Family is a connected group of artificial intelligence technologies, systems, behaviors, and applications that work together to help machines perform intelligent tasks. This ecosystem includes core AI systems, AI agents, agentic AI, AI autonomy, autonomous AI workers, and generative AI. Each part plays a different role in how modern AI systems think, learn, create, make decisions, solve problems, and interact with the world.

At the center of the ecosystem is artificial intelligence itself — the broad field focused on building machines and software that can perform tasks that normally require human intelligence. AI agents act as the systems that carry out goals and actions. Agentic AI describes the ability of AI systems to plan, reason, adapt, and pursue objectives with limited human guidance. AI autonomy represents the level of independence an AI system has while making decisions and taking actions. AI autonomous workers apply these capabilities to perform real-world tasks like digital employees or assistants. Generative AI adds the ability to create new content such as text, images, music, video, code, and designs.

The AI Ecosystem Family demonstrates that modern artificial intelligence is no longer just one single technology. Instead, it is an interconnected environment of systems, tools, intelligence, automation, creativity, and human collaboration. As AI continues to evolve, these technologies are becoming increasingly integrated into education, business, healthcare, communication, cybersecurity, research, and everyday life.

Understanding the AI Ecosystem Family helps students, educators, and readers see how the different parts of AI connect together to shape the future of intelligent systems and the digital world.

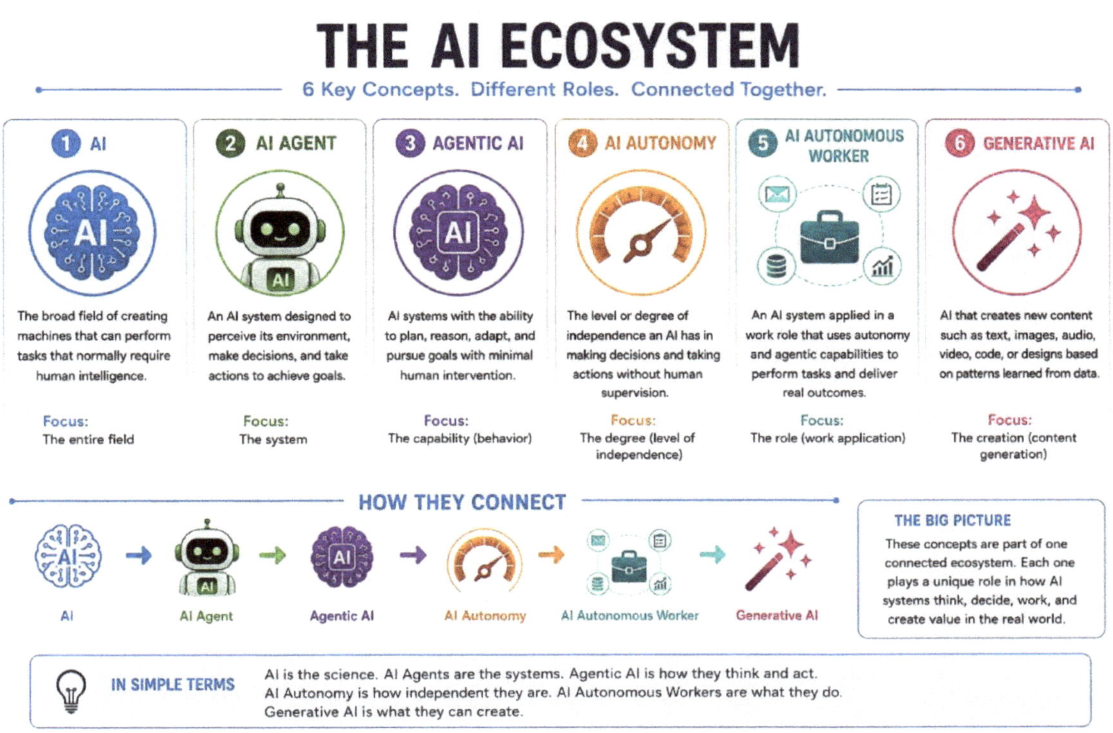

Parameter Family Introduction

The Parameter Family represents the internal learning system of artificial intelligence. While tokens provide the language and data provides the learning material, parameters are what allow AI systems to adjust, improve, and make intelligent decisions over time.

In simple terms, parameters are numerical values inside an AI model that change as the system learns. These values are not visible to users, but they play a critical role in how the model understands patterns, relationships, and meaning. Every time an AI system is trained, it adjusts its parameters to better predict outcomes and generate accurate responses.

Parameters exist throughout neural networks, where they act as the connections between artificial neurons. Each connection has a weight, which determines how important a particular piece of information is. During training, these weights are continuously adjusted based on data, allowing the system to improve its performance.

The Parameter Family works closely with the Data Family, which provides the information needed for learning, and the Neural Network Family, which organizes how parameters are structured and applied. It also supports the Token Family by helping the model interpret and predict sequences of language.

Understanding parameters helps students recognize that AI is not simply programmed with fixed rules. Instead, it learns by adjusting internal values that guide how it processes information. These adjustments are what allow AI systems to evolve from simple pattern recognition to complex reasoning and decision-making.

In essence, if data is what AI learns from, parameters are how AI learns.

Parameter Family Breakdown Chart

Component	Role	Description	Example
Parameters	Learning Values	Numerical values inside an AI model that are adjusted during training	Millions or billions of values inside a model
Weights	Connection Strength	Values that determine how strongly one neuron influences another	A higher weight = stronger influence
Bias	Adjustment Factor	A value added to help the model shift and fine-tune outputs	Helps improve accuracy in predictions
Training Process	Learning Mechanism	The process of adjusting parameters using data to improve performance	Model improves after analyzing many examples

Component	Role	Description	Example
Loss Function	Error Measurement	Measures how far the model's prediction is from the correct answer	Lower loss = better performance
Optimization Algorithm	Adjustment Strategy	Method used to update parameters efficiently	Gradient Descent adjusts weights step by step
Gradient	Direction of Change	Indicates how parameters should change to reduce error	Shows whether to increase or decrease a value
Backpropagation	Learning Process	System that sends error signals backward to update parameters	Adjusts earlier layers based on output error
Hyperparameters	Control Settings	External settings that guide how training happens (not learned)	Learning rate, batch size
Model Size	Capacity Indicator	The total number of parameters in a model	Larger models = more learning capacity

The Parameter Family is the hidden engine of learning in artificial intelligence. While users interact with outputs, it is the continuous adjustment of parameters behind the scenes that makes intelligent behavior possible.

Data Family Introduction

The Data Family represents the foundation of learning in artificial intelligence. Every AI system depends on data to develop knowledge, recognize patterns, and make decisions. Without data, artificial intelligence cannot learn, adapt, or function effectively.

Data can take many forms, including text, images, audio, video, and numerical information. These inputs provide the raw material that AI systems analyze during training. By studying large amounts of data, AI models begin to identify patterns, relationships, and structures that allow them to make predictions and generate responses.

The quality and diversity of data play a critical role in how well an AI system performs. Accurate, balanced, and relevant data helps models produce reliable results, while poor or biased data can lead to incorrect or unfair outcomes. This is why data collection, preparation, and evaluation are essential steps in the development of responsible AI systems.

The Data Family works closely with the Parameter Family, which adjusts internal values based on data, and the Neural Network Family, which processes and organizes data through layers of computation. It also supports the Token Family when dealing with language, as text data is converted into tokens for analysis.

Understanding the Data Family helps students recognize that AI does not "know" things on its own. Instead, it learns from the information it is given. The more meaningful and well-prepared the data, the more capable the AI system becomes.

In simple terms, data is the source of knowledge, and it is the starting point for all artificial intelligence.

Data Family Breakdown Chart

Component	Role	Description	Example
Data	Learning Material	Raw information used by AI systems to learn patterns and relationships	Text, images, audio, numbers
Dataset	Organized Collection	A structured group of data used for training and evaluation	A folder of labeled images
Training Data	Learning Input	Data used to teach the AI model during training	Thousands of sentences for a language model
Validation Data	Performance Check	Data used to tune the model during training without directly learning from it	Helps prevent overfitting
Test Data	Final Evaluation	Data used to measure how well the model performs after training	New unseen examples
Structured Data	Organized Format	Data arranged in tables with clear categories	Spreadsheets, databases
Unstructured Data	Flexible Format	Data without a predefined structure	Text documents, images, videos
Labeled Data	Guided Learning	Data tagged with correct answers to guide training	Image labeled "cat" or "dog"
Unlabeled Data	Self-Learning Input	Data without labels, used in unsupervised learning	Raw text without categories
Data Quality	Accuracy Measure	The reliability and correctness of data	Clean, complete, and consistent data
Data Bias	Imbalance Issue	When data does not represent all groups fairly	Skewed or incomplete datasets
Data Preprocessing	Preparation Step	Cleaning and organizing data before training	Removing errors, formatting text

The Data Family reminds us that artificial intelligence is only as strong as the information it learns from. High-quality data leads to meaningful insights, while poor data can limit or misguide intelligent systems.

Memory Family Introduction

The Memory Family represents how artificial intelligence systems store, retain, and retrieve information. Just as humans rely on memory to recall past experiences, learn new concepts, and make decisions, AI systems depend on memory to access previously processed information and maintain context.

In artificial intelligence, memory is not a single component but a collection of mechanisms that allow systems to handle information over time. Some forms of memory are short-term, holding information temporarily while a task is being completed. Other forms are long-term, storing learned knowledge that can be used across different tasks and interactions.

Memory plays a critical role in enabling AI systems to understand sequences, maintain context in conversations, and improve decision-making. For example, when a user interacts with a language model, the system uses memory to keep track of previous words, sentences, or prompts in order to generate coherent and relevant responses.

The Memory Family works closely with the Data Family, which provides the information to be stored, and the Parameter Family, which encodes learned knowledge within the model. It also supports the Neural Network Family by enabling systems to process sequences and retain important patterns over time.

Understanding the Memory Family helps students see that AI is not simply reacting to individual inputs in isolation. Instead, it relies on stored information and contextual awareness to produce meaningful and consistent outputs.

In simple terms, if data is what AI learns from, memory is how AI remembers and uses what it has learned.

Memory Family Breakdown Chart

Component	Role	Description	Example
Memory	Information Storage	The ability of an AI system to store and access information	Retaining previous inputs in a task
Short-Term Memory	Temporary Storage	Holds information for immediate processing	Keeping track of words in a sentence
Long-Term Memory	Persistent Storage	Stores learned knowledge for future use	Knowledge embedded in model parameters
Context Memory	Conversation Tracking	Maintains information within a specific interaction	Remembering earlier parts of a prompt
Working Memory	Active Processing	Handles information currently being used in computation	Processing a sequence step by step

Component	Role	Description	Example
External Memory	Extended Storage	Memory stored outside the model for retrieval	Databases, vector stores
Internal Memory	Embedded Knowledge	Information stored within the model itself	Learned patterns inside neural networks
Sequence Memory	Order Awareness	Tracks the order of inputs over time	Understanding sentence structure
Retrieval Mechanism	Access System	Retrieves stored information when needed	Searching relevant past data
Attention Mechanism	Focus System	Highlights important parts of memory for processing	Focusing on key words in a sentence
Memory Capacity	Storage Limit	The amount of information the system can handle at once	Context window size

The Memory Family shows that intelligence is not just about processing information in the moment, but about retaining, organizing, and using knowledge over time to create meaningful understanding.

Knowledge Family Introduction

The Knowledge Family represents how artificial intelligence organizes, connects, and uses information to create understanding. While data provides raw information and memory stores it, knowledge is what gives that information meaning and structure.

In artificial intelligence, knowledge is not simply a collection of facts. It is the result of identifying relationships, patterns, and connections within data. AI systems use knowledge to recognize concepts, make decisions, and generate meaningful responses. This allows machines to move beyond simple data processing and toward more intelligent behavior.

Knowledge can be represented in different ways within AI systems. Some forms are structured, such as databases and knowledge graphs, where relationships between concepts are clearly defined. Other forms are learned implicitly through neural networks, where patterns are encoded within parameters and used to guide predictions.

The Knowledge Family works closely with the Data Family, which provides the information needed to build knowledge, and the Memory Family, which stores and retrieves that information. It also connects with the Parameter Family, where learned knowledge is embedded within the model, and the Neural Network Family, which processes and organizes that knowledge.

Understanding the Knowledge Family helps students recognize that AI systems do not simply memorize information. Instead, they learn how pieces of information relate to one another, allowing them to interpret meaning, solve problems, and respond intelligently.

In simple terms, if data is information and memory stores it, knowledge is the understanding that makes it useful.

Knowledge Family Breakdown Chart

Component	Role	Description	Example
Knowledge	Understanding System	Organized information that allows AI to interpret meaning and make decisions	Recognizing relationships between concepts
Facts	Basic Information	Individual pieces of information stored within a system	"Water freezes at 0°C"
Relationships	Connection Mapping	Links between pieces of information that create meaning	"Teacher teaches student"
Knowledge Representation	Structure System	Methods used to organize and store knowledge in AI	Knowledge graphs, semantic networks
Knowledge Graph	Network Mapping	A structured representation of entities and their relationships	Google Knowledge Graph
Semantic Understanding	Meaning Interpretation	The ability to understand meaning behind words and data	Understanding synonyms and context
Inference	Reasoning Process	Drawing conclusions based on known information	If A = B and B = C, then A = C
Rules	Decision Framework	Logical guidelines used to make decisions	If temperature < 0°C → freeze
Ontology	Concept Framework	A structured system defining categories and relationships	Classification of animals
Explicit Knowledge	Direct Information	Clearly defined and stored knowledge	Rules, databases
Implicit Knowledge	Learned Patterns	Knowledge learned through experience and data	Patterns learned by neural networks
Knowledge Integration	System Coordination	Combining information from multiple sources	Merging text, images, and data

The Knowledge Family transforms information into understanding, allowing artificial intelligence to move beyond data and memory into meaningful reasoning and intelligent decision-making.

AI Model Family Introduction

The AI Model Family represents the core system within artificial intelligence that performs tasks such as prediction, classification, and content generation. It is the part of AI that users interact with, whether they are asking questions, generating images, or analyzing data.

An AI model is a trained system that has learned patterns from data. During training, the model processes large amounts of information and adjusts its internal parameters to improve accuracy. Once trained, the model can take new input and produce meaningful output based on what it has learned.

AI models can take many forms depending on their purpose. Some models are designed to recognize images, others to understand language, and others to make decisions or predictions. Large Language Models (LLMs), for example, are designed to process and generate human language by predicting sequences of tokens.

The AI Model Family works closely with the Data Family, which provides the learning material, and the Parameter Family, which stores learned patterns. It also relies on the Neural Network Family for structure and the Chip and Infrastructure Families for execution and deployment.

Understanding the AI Model Family helps students see that AI is not just a concept, but a working system that has been trained to perform specific tasks. It is the engine that transforms input into output.

In simple terms, if AI is the system, the model is the part that actually does the thinking and producing.

AI Model Family Breakdown Chart

Component	Role	Description	Example
AI Model	Core System	The trained system that performs tasks	Chatbot, image generator
Training	Learning Process	Process of teaching the model using data	Learning from datasets
Inference	Output Generation	Using the trained model to produce results	Answering a question
Model Architecture	Structural Design	Defines how the model is built	Neural networks, transformers
Parameters	Learned Values	Internal values adjusted during training	Billions of weights
Input Data	Entry Information	Data provided to the model for processing	User prompt
Output Data	Result	The response or prediction generated	AI-generated text
Model Types	Functional Categories	Different kinds of models for different tasks	Classification, generation
Fine-Tuning	Model Adjustment	Improving a model for specific tasks	Custom-trained AI

Component	Role	Description	Example
Pretrained Model	Base System	A model trained on large datasets before use	General-purpose AI
Evaluation	Performance Check	Measuring how well the model performs	Accuracy, loss metrics

The AI Model Family represents the working core of artificial intelligence, where learned patterns are transformed into meaningful actions, predictions, and responses.

Neural Network Family Introduction

The Neural Network Family represents the structural foundation of how artificial intelligence learns and processes information. Inspired by the human brain, neural networks are systems of interconnected units, often called neurons, that work together to analyze data, recognize patterns, and make decisions.

In artificial intelligence, a neural network is made up of layers. The input layer receives data, the hidden layers process that data, and the output layer produces a result. As information flows through these layers, the network applies mathematical transformations that allow it to detect patterns and relationships within the data.

Neural networks rely heavily on the Parameter Family, where weights and biases determine how strongly information flows between neurons. During training, these parameters are adjusted to improve the network's performance. The more the network learns, the better it becomes at recognizing patterns and making accurate predictions.

The Neural Network Family works closely with the Data Family, which provides the information needed for learning, and the Memory and Knowledge Families, which help store and organize learned information. It also supports the Token Family when processing language, allowing AI systems to interpret sequences of tokens in meaningful ways.

Understanding neural networks helps students see that AI is not simply following instructions. Instead, it is learning from examples and improving over time through layered processing and pattern recognition.

In simple terms, neural networks are the systems that allow AI to think, learn, and make sense of information.

Neural Network Family Breakdown Chart

Component	Role	Description	Example
Neural Network	Learning Structure	A system of interconnected neurons that processes data and learns patterns	Image recognition system
Neuron (Node)	Processing Unit	A basic unit that receives input, processes it, and passes output forward	A single calculation point
Input Layer	Data Entry Point	The first layer that receives raw data	Pixels of an image
Hidden Layers	Processing Layers	Intermediate layers that transform data and detect patterns	Feature detection layers
Output Layer	Result Generator	The final layer that produces predictions or decisions	"Cat" vs "Dog" classification
Weights	Signal Strength	Values that determine how strongly inputs influence outputs	Higher weight = stronger impact
Bias	Adjustment Value	Helps shift outputs to improve learning accuracy	Fine-tuning predictions
Activation Function	Decision Function	Determines whether a neuron should activate based on input	ReLU, Sigmoid
Forward Propagation	Data Flow	The process of passing input data through the network to produce output	Input → Hidden → Output
Backpropagation	Learning Process	Adjusts weights and biases based on error to improve performance	Correcting mistakes
Loss Function	Error Measurement	Calculates how far the output is from the correct answer	Prediction vs actual
Deep Neural Network	Advanced Structure	A network with many hidden layers for complex learning	Deep learning models

The Neural Network Family provides the structure that allows artificial intelligence to transform data into understanding, making it possible for machines to learn from experience and improve over time.

Dual Language System in Artificial Intelligence

Definition

The Dual Language System in Artificial Intelligence refers to the way AI communicates using two different forms of language at the same time: human language for interaction and machine language for internal processing.

Key Components

Human Language (Natural Language)
This is the language people use to communicate with AI, such as English, Spanish, Tongan, or Fijian. It includes words, sentences, and questions that users type or speak.

Machine Language (Numerical Representation)
This is how AI understands information internally. Words are converted into numbers called tokens and processed using mathematical models and algorithms.

How It Works (Step-by-Step)

A user enters a question in human language

The AI converts the words into numerical tokens

The AI processes the tokens using a neural network

The AI generates a response based on patterns and probabilities

The response is converted back into human language

Example

User Input:
"What is Artificial Intelligence?"

AI Process:

Converts words into numbers

Analyzes patterns from training data

Predicts the best possible answer

Output:
"Artificial Intelligence is the ability of machines to perform tasks that normally require human intelligence."

Why This Matters

Understanding the dual language system helps students realize that:

AI does not "think" like humans

AI translates language into math and back into language

Every interaction with AI is based on patterns, not emotions or understanding

Applications

Chatbots and virtual assistants

Language translation systems

Search engines

Voice recognition tools

Challenges

Misinterpretation of human language

Bias in language data

Lack of true understanding or reasoning

Summary

Artificial Intelligence works by translating human language into machine-readable data and then converting the results back into human language. This dual system allows AI to interact with people while operating through mathematical processes behind the scenes.

AI Translator Architecture Family Introduction

The AI Translator Architecture Family represents the system that transforms input into meaningful output in artificial intelligence. It acts as the bridge between what humans provide and what AI produces. Whether a user types a question, uploads an image, or speaks a command, this architecture is responsible for interpreting that input and generating a response.

In artificial intelligence, translation does not only mean converting one language to another. It refers to the broader process of transforming one form of information into another. For example, text can be translated into a response, speech can be converted into text, and images can be interpreted into descriptions. This transformation process is at the core of how modern AI systems function.

The AI Translator Architecture is built on multiple foundational families. It relies on the Token Family to break input into manageable units, the Neural Network Family to process patterns, and the Parameter Family to guide learning. It also uses knowledge and memory to maintain context and produce coherent results.

At the center of this architecture is the ability to encode and decode information. Encoding transforms input into a format the model can understand, while decoding converts processed information back into human-readable output. This continuous transformation allows AI systems to interact with users in a natural and meaningful way.

Understanding the AI Translator Architecture Family helps students see that AI is not simply responding randomly. Instead, it is systematically translating input into output through structured processes and learned patterns.

In simple terms, if tokens are the language pieces and neural networks are the brain, the AI Translator Architecture is the system that turns understanding into communication.

AI Translator Architecture Family Breakdown Chart

Component	Role	Description	Example
AI Translator Architecture	Transformation System	Converts input into meaningful output	Question → Answer
Input Processing	Data Intake	Receives and prepares user input for analysis	Text prompt entered by a user
Encoding	Input Conversion	Transforms input into numerical representations	Words → token IDs
Context Building	Meaning Formation	Organizes input into a structured understanding	Sentence context tracking
Sequence Processing	Order Handling	Maintains the correct order of information	Word sequence in a sentence
Attention Mechanism	Focus System	Identifies important parts of input data	Highlighting key words
Representation Layer	Feature Mapping	Converts data into internal representations for processing	Embeddings
Decoding	Output Generation	Converts processed data back into human-readable form	Token IDs → words
Output Generation	Response Creation	Produces the final answer or result	AI-generated response
Multimodal Translation	Cross-Format Processing	Translates between different types of data	Image → text description
Feedback Loop	Improvement Cycle	Uses results to refine future outputs	Model learning over time

The AI Translator Architecture Family shows that artificial intelligence is not just about understanding information, but about transforming it into meaningful communication that humans can use and interact with.

AI Architecture Family Introduction

The AI Architecture Family represents the overall design and structure of an artificial intelligence system. While individual families such as Data, Parameters, and Neural Networks explain how AI learns and processes information, the AI Architecture Family shows how all these components are organized and work together as a complete system.

In artificial intelligence, architecture refers to the blueprint that defines how data flows through a system, how components interact, and how decisions are produced. It determines how input is received, how it is processed, and how output is generated. A well-designed architecture allows AI systems to operate efficiently, scale to large problems, and produce reliable results.

AI architectures can vary depending on the type of system being built. Some are designed for image recognition, others for language processing, and others for decision-making or automation. Despite these differences, most architectures share common elements such as input layers, processing layers, memory components, and output mechanisms.

The AI Architecture Family connects all foundational families. It integrates the Data Family for input, the Token Family for language processing, the Neural Network Family for computation, the Parameter Family for learning, and the Memory and Knowledge Families for storing and organizing information. It also works closely with the AI Translator Architecture Family to transform input into meaningful output.

Understanding the AI Architecture Family helps students see the "big picture" of artificial intelligence. Instead of viewing AI as separate parts, they begin to understand it as a coordinated system where each component plays a specific role.

In simple terms, if individual families are the parts of AI, the AI Architecture Family is the blueprint that brings everything together into one working system.

AI Architecture Family Breakdown Chart

Component	Role	Description	Example
AI Architecture	System Blueprint	The overall design that defines how an AI system is structured and operates	ChatGPT system design

Component	Role	Description	Example
Input Layer	Data Entry Point	Receives raw input from users or external sources	Text prompt, image upload
Preprocessing Layer	Data Preparation	Cleans and organizes input data before processing	Tokenization, normalization
Processing Core	Computation Engine	The main system where data is analyzed and transformed	Neural networks, transformers
Parameter System	Learning Core	Stores adjustable values that guide learning and predictions	Model weights and biases
Memory System	Context Storage	Maintains and retrieves information during processing	Context window, external memory
Knowledge System	Understanding Layer	Organizes relationships and meaning within the system	Knowledge graphs, learned patterns
Decision Layer	Output Logic	Determines the final output based on processed data	Selecting the best response
Output Layer	Result Delivery	Presents results to the user in a usable format	Text, image, or audio response
Feedback Loop	Improvement System	Uses outcomes to refine future performance	Model updates and retraining
Scalability System	Expansion Capability	Allows the system to handle increasing data and complexity	Cloud-based AI systems
Integration Layer	System Connection	Connects AI with external tools and platforms	APIs, databases, applications

The AI Architecture Family reveals that artificial intelligence is not a single technology, but a coordinated system of interconnected components working together to transform data into intelligent outcomes.

Chip Family Introduction

The Chip Family represents the physical hardware that powers artificial intelligence systems. While software components such as data, algorithms, and neural networks define how AI works, none of these systems can function without the computing chips that perform the actual calculations.

Chips are specialized electronic components designed to process information at extremely high speeds. In artificial intelligence, these chips handle the complex mathematical operations required for training models and generating outputs. Every prediction, calculation, and response produced by an AI system is executed by hardware within the Chip Family.

Different types of chips are used depending on the needs of the system. Central Processing Units (CPUs) handle general-purpose tasks, while Graphics Processing Units (GPUs) are optimized for parallel processing, making them ideal for training neural networks. More specialized chips, such as Tensor Processing Units (TPUs) and Neural Processing Units (NPUs), are designed specifically for AI workloads, enabling faster and more efficient computation.

The Chip Family works closely with the Neural Network Family, which defines the structure of computation, and the Parameter Family, which stores the values being processed. It also connects to the Infrastructure Family, where large-scale systems of chips are organized in data centers and cloud environments.

Understanding the Chip Family helps students recognize that artificial intelligence is not purely abstract. It depends on physical machines that require power, resources, and engineering to operate. These chips are the engines that bring AI models to life.

In simple terms, if AI is the intelligence, chips are the machines that make that intelligence possible.

Chip Family Breakdown Chart

Component	Role	Description	Example
Chip (Processor)	Computation Engine	Electronic hardware that performs calculations required for AI	CPU, GPU
CPU (Central Processing Unit)	General Processor	Handles a wide range of computing tasks	Running basic applications
GPU (Graphics Processing Unit)	Parallel Processor	Processes many calculations at once, ideal for AI training	Training neural networks
TPU (Tensor Processing Unit)	AI Accelerator	Specialized chip designed for machine learning tasks	Google TPU
NPU (Neural Processing Unit)	Edge AI Processor	Optimized for running AI on devices like phones	Smartphone AI features
AI Accelerator	Performance Booster	Hardware designed to speed up AI computations	Dedicated AI chips
Memory (Hardware)	Data Storage	Stores data and instructions for processing	RAM, VRAM
Parallel Processing	Speed Mechanism	Ability to perform multiple calculations simultaneously	GPUs processing thousands of operations
Throughput	Processing Capacity	Amount of data processed over time	High-performance computing systems
Latency	Response Time	Time it takes to process a request	Faster chips = lower latency
Power Consumption	Energy Usage	Amount of energy required to run the chip	High-performance GPUs use more power

Component	Role	Description	Example
Edge Devices	Local Processing	Devices that run AI locally without cloud support	Phones, smart cameras

The Chip Family reminds us that behind every intelligent system is powerful hardware performing millions or billions of calculations, turning abstract models into real-world functionality.

Watts Family Introduction

The Watts Family represents the measurement of power used by artificial intelligence systems. While the Chip Family provides the hardware that performs computations, the Watts Family explains how much energy those systems consume while operating.

A watt is a unit of power that measures the rate at which energy is used. In artificial intelligence, watts help us understand how much electricity is required to run processors, train models, and generate responses. Every time an AI system processes data, performs calculations, or produces output, it consumes power.

Different AI systems require different levels of power depending on their size and complexity. Small systems, such as those running on mobile devices, use relatively low amounts of power. In contrast, large-scale AI models operating in data centers can require significant amounts of electricity, especially during training, where billions of calculations are performed continuously.

The Watts Family works closely with the Chip Family, which determines how efficiently computations are performed, and the Infrastructure Family, where large networks of machines operate together. It also connects to the overall design of AI systems, as more efficient models can reduce power consumption while maintaining performance.

Understanding the Watts Family helps students recognize that artificial intelligence is not only about intelligence and computation, but also about energy use and resource management. As AI continues to grow, managing power efficiently becomes an important part of building responsible and sustainable systems.

In simple terms, if chips perform the work, watts measure how much power it takes to do that work.

Watts Family Breakdown Chart

Component	Role	Description	Example
Watt (W)	Power Measurement	Unit that measures the rate of energy use	A device using 100 watts
Power Consumption	Energy Usage Rate	Amount of power used during operation	AI model running on GPUs

Component	Role	Description	Example
Energy Efficiency	Performance Balance	How effectively a system uses power to perform tasks	More output with less power
High-Performance Computing	Intensive Power Use	Systems that require large amounts of power for complex tasks	AI training clusters
Idle Power	Standby Usage	Power consumed when systems are not actively processing	Servers waiting for requests
Peak Power	Maximum Usage	Highest level of power used during heavy workloads	Training large AI models
Thermal Output	Heat Generation	Heat produced as a result of power usage	Cooling systems in data centers
Cooling Systems	Temperature Control	Systems used to manage heat from high power usage	Air or liquid cooling
Power Supply	Energy Source	Provides electricity to AI systems	Electrical grid, batteries
Edge Power Usage	Local Energy Use	Power consumption on smaller devices	Smartphones, IoT devices
Data Center Power	Large-Scale Usage	Power required for large AI operations	Server farms running AI models
Sustainable Energy	Efficiency Goal	Use of renewable energy to reduce environmental impact	Solar-powered data centers

The Watts Family highlights that intelligence comes with a cost—every AI system requires power, and understanding that power is essential for building efficient and sustainable technologies.

Watts Family (AI Power Usage Scale)

Definition

The **Watts Family** represents the real-time power usage of artificial intelligence systems, measured in watts (W). It shows how much power AI systems consume at a given moment as they process data and perform computations.

Watts Scale

1 Watt
10 Watts
100 Watts
500 Watts
1,000 Watts (1 Kilowatt)
10,000 Watts (10 Kilowatts)

100,000 Watts (100 Kilowatts)
1,000,000 Watts (1 Megawatt)
10,000,000 Watts (10 Megawatts)
100,000,000+ Watts (100+ Megawatts)

Key Characteristics

Measured in real-world electrical power

Higher watt values indicate higher processing demand

Used to represent the scale of AI systems from small devices to large infrastructures

Closely connected to the Chip Family, Energy Family, and Infrastructure Family

Examples

Low watts → small devices and basic AI functions

Medium watts → computers and advanced applications

High watts → data centers and large AI systems

Summary

The **Watts Family** provides a numerical representation of how much power artificial intelligence systems use in real time, helping to illustrate the scale and intensity of AI operations.

🧠 Student Insight

Watts show how much power AI is using right now.

AI Energy Family Introduction

The AI Energy Family represents the total amount of energy consumed by artificial intelligence systems over time. While the Watts Family measures the rate at which power is used at any given moment, the Energy Family focuses on the accumulated energy required to perform tasks, run systems, and sustain operations.

In artificial intelligence, energy is used every time a model is trained, a system processes data, or a response is generated. Large-scale AI systems, especially those operating in data centers, can consume significant amounts of energy due to the continuous processing of massive datasets and complex computations.

Energy is typically measured over time using units such as kilowatt-hours (kWh), which reflect how much power has been used during a specific period. For example, running a high-performance AI system for several hours or days results in a total energy cost that goes beyond the moment-to-moment power usage measured in watts.

The AI Energy Family works closely with the Watts Family, which provides the rate of power consumption, and the Chip Family, which determines how efficiently computations are performed. It also connects to the Infrastructure Family, where large-scale systems operate continuously and require long-term energy management.

Understanding the AI Energy Family helps students recognize that artificial intelligence has real-world resource implications. It is not only about performance and speed, but also about sustainability, efficiency, and responsible use of technology.

In simple terms, if watts measure how fast energy is used, energy measures how much is used over time.

AI Energy Family Breakdown Chart

Component	Role	Description	Example
Energy	Total Consumption	The total amount of power used over a period of time	Running an AI system for hours or days
Kilowatt-hour (kWh)	Energy Measurement	Unit used to measure energy consumption over time	1 kWh = using 1,000 watts for 1 hour
Energy Usage	Consumption Tracking	Total energy required for AI operations	Training a large model
Training Energy	Learning Cost	Energy used during model training processes	Weeks of GPU usage
Inference Energy	Response Cost	Energy used when AI generates outputs	Answering a user prompt
Energy Efficiency	Optimization Goal	Reducing energy use while maintaining performance	Efficient AI models
Energy Scaling	Growth Impact	Increase in energy use as systems grow larger	Bigger models = more energy
Data Center Energy	Infrastructure Demand	Total energy used by large AI facilities	Server farms operating 24/7
Cooling Energy	Temperature Control	Energy required to cool systems and prevent overheating	Air conditioning in data centers
Renewable Energy	Sustainability Source	Use of clean energy to power AI systems	Solar, wind-powered data centers

Component	Role	Description	Example
Carbon Impact	Environmental Effect	Emissions associated with energy consumption	AI training carbon footprint
Energy Management	Resource Control	Strategies to monitor and reduce energy use	Efficient scheduling, hardware optimization

The AI Energy Family reminds us that artificial intelligence operates within the physical world, where every computation consumes energy and every system has an impact on resources and sustainability.

Energy Family (Total Power Consumption of Artificial Intelligence)

Definition

The **Energy Family** represents the total amount of electricity used by artificial intelligence systems over time. It is measured in watt-hours (Wh) or kilowatt-hours (kWh) and shows how much energy AI systems consume while running.

Energy Scale

1 Watt-hour (Wh)
10 Watt-hours (Wh)
100 Watt-hours (Wh)
500 Watt-hours (Wh)
1,000 Watt-hours (1 Kilowatt-hour, kWh)
10,000 Watt-hours (10 kWh)
100,000 Watt-hours (100 kWh)
1,000,000 Watt-hours (1 Megawatt-hour, MWh)
10,000,000 Watt-hours (10 MWh)
100,000,000+ Watt-hours (100+ MWh)

Key Characteristics

Measured over time (not instant like watts)

Represents total electricity consumption

Increases the longer AI systems run

Directly related to cost and energy usage

Closely connected to Watts Family, Chip Family, and Infrastructure Family

Examples

Low energy → short AI tasks or small devices

Medium energy → daily AI usage on computers

High energy → training AI models and running data centers

Summary

The **Energy Family** provides a numerical representation of the total electricity consumed by artificial intelligence systems over time, helping to illustrate the overall cost and impact of AI operations.

🧠 Student Insight

Energy shows how much total power AI has used over time.

AI Infrastructure Family Introduction

The AI Infrastructure Family represents the large-scale systems and environments that support, power, and connect artificial intelligence technologies. While individual components such as chips, data, and models explain how AI works, infrastructure explains where and how these systems operate in the real world.

AI infrastructure includes data centers, cloud platforms, networking systems, and storage environments that allow artificial intelligence to function at scale. These systems provide the computing power, data access, and connectivity needed to train models, process information, and deliver results to users around the world.

Modern AI systems rely heavily on cloud-based infrastructure, where thousands of machines work together to handle massive workloads. This allows AI to scale efficiently, making it possible to process large datasets, run complex models, and serve millions of users simultaneously. Infrastructure also ensures reliability, security, and continuous availability of AI services.

The AI Infrastructure Family works closely with the Chip Family, which provides the hardware, the Watts and Energy Families, which manage power consumption, and the Data Family, which supplies the information being processed. It also supports the AI Architecture Family by providing the environment where all system components are deployed and connected.

Understanding the AI Infrastructure Family helps students recognize that artificial intelligence is not contained within a single device. Instead, it operates across global systems that require coordination, resources, and engineering to function effectively.

In simple terms, if AI is the system and chips are the engines, infrastructure is the environment that allows everything to run at scale.

AI Infrastructure Family Breakdown Chart

Component	Role	Description	Example
AI Infrastructure	System Environment	The physical and digital systems that support AI operations	Global AI platforms
Data Center	Processing Hub	Facilities that house servers and computing equipment	Large server farms
Cloud Computing	Scalable Platform	Remote systems that provide computing resources over the internet	AWS, Azure, Google Cloud
Servers	Compute Units	Machines that process data and run AI models	Rack-mounted servers
Networking	Connectivity System	Systems that allow communication between machines	Internet, fiber networks
Storage Systems	Data Management	Systems that store large volumes of data	Databases, cloud storage
Distributed Computing	Workload Sharing	Splitting tasks across multiple machines for efficiency	Parallel processing across clusters
Edge Infrastructure	Local Processing	Running AI closer to the user or device	Smart devices, IoT systems
Load Balancing	Traffic Control	Distributes workloads to prevent overload	Managing user requests
Security Systems	Protection Layer	Safeguards data and systems from threats	Encryption, firewalls
Redundancy	Reliability System	Backup systems to ensure continuous operation	Failover servers
Scalability	Growth Capability	Ability to expand resources as demand increases	Adding more servers dynamically

The AI Infrastructure Family reveals that artificial intelligence operates on a global scale, relying on interconnected systems that provide the power, storage, and connectivity needed to support intelligent technologies.

AI Agent Family Introduction

The AI Agent Family represents systems that can take action, make decisions, and perform tasks on behalf of users or other systems. Unlike traditional AI models that only respond to input, AI agents are designed to operate with a level of autonomy, allowing them to plan, execute, and adapt to achieve specific goals.

An AI agent receives input from its environment, processes that information, and takes actions based on predefined objectives or learned behaviors. These actions can include answering questions, automating workflows, retrieving information, or interacting with other systems. Some agents operate in simple environments, while others function in complex systems that require continuous decision-making.

AI agents often combine multiple foundational families. They use the Data Family to gather information, the Memory and Knowledge Families to retain and understand context, and the Neural Network and Parameter Families to process and learn from interactions. They also rely on the AI Architecture and Infrastructure Families to operate reliably at scale.

Modern AI agents can work independently or as part of larger systems. For example, a customer support agent can respond to inquiries automatically, while a more advanced agent can complete multi-step tasks such as scheduling, research, or system management.

Understanding the AI Agent Family helps students see that artificial intelligence is not only about generating responses, but also about taking meaningful action in real-world scenarios.

In simple terms, if AI models think and respond, AI agents act and execute.

AI Agent Family Breakdown Chart

Component	Role	Description	Example
AI Agent	Action System	An AI system that can make decisions and perform tasks	Virtual assistant
Environment	Operating Context	The space where the agent interacts and gathers information	Web, apps, databases
Input Perception	Data Intake	Receives information from the environment	User query or sensor input
Decision Engine	Action Selection	Determines what action to take based on input	Choosing a response or task
Action Execution	Task Performance	Carries out decisions in the environment	Sending an email, retrieving data
Goal System	Objective Setting	Defines what the agent is trying to achieve	Complete a task or solve a problem
Planning Module	Strategy Builder	Breaks down tasks into steps	Multi-step problem solving

Component	Role	Description	Example
Feedback Loop	Learning Cycle	Improves performance based on outcomes	Adjusting actions over time
Autonomy Level	Independence Scale	Degree of independence in decision-making	Fully automated vs assisted
Multi-Agent System	Collaboration Network	Multiple agents working together	Coordinated AI systems

Ingredient Family Introduction

The Ingredient Family represents the essential components that come together to create an artificial intelligence system. Just as a recipe requires specific ingredients to produce a final dish, AI systems rely on a combination of foundational elements working together to function effectively.

Artificial intelligence is not built from a single component. It requires data to learn from, algorithms to guide processing, models to perform tasks, hardware to execute computations, and energy to power the entire system. Each of these elements plays a critical role, and the absence of any one component would prevent the system from operating properly.

The Ingredient Family provides a simplified way for students to understand AI as a system made up of interconnected parts. Instead of viewing AI as complex or mysterious, learners can begin to see it as something constructed from identifiable and understandable components.

This family connects directly with all other foundational families, including Data, Algorithm, Neural Network, Parameter, Chip, and Energy Families. It serves as an overview that brings these elements together into a unified perspective.

Understanding the Ingredient Family helps students recognize that artificial intelligence is built, not magical. It is the result of carefully combining the right components in the right way.

In simple terms, if AI is the final product, the Ingredient Family represents everything needed to build it.

Ingredient Family Breakdown Chart

Component	Role	Description	Example
Data	Learning Input	Provides the information AI uses to learn patterns	Text, images, audio
Algorithms	Instruction Set	Guides how data is processed and decisions are made	Sorting, classification
AI Models	Execution System	Performs tasks such as prediction or generation	Language models

Component	Role	Description	Example
Neural Networks	Processing Structure	Organizes how data is processed through layers	Deep learning models
Parameters	Learning Adjustments	Fine-tune how the model learns and makes predictions	Weights and biases
Tokens	Language Units	Breaks text into smaller pieces for processing	Words or subwords
Hardware (Chips)	Computation Engine	Executes calculations required by AI systems	GPUs, CPUs
Energy	Power Source	Supplies the energy needed to run systems	Electricity usage
Infrastructure	Support System	Provides the environment for AI to operate at scale	Cloud systems, data centers
Memory	Storage System	Stores and retrieves information for processing	Context memory
Knowledge	Understanding Layer	Represents meaning and relationships within data	Knowledge graphs

The Ingredient Family shows that artificial intelligence is built from a combination of essential components, each playing a vital role in creating intelligent systems.

The Frontend and Backend Family of Artificial Intelligence

How AI Systems Connect User Interaction to Intelligent Processing

Introduction

Every artificial intelligence system operates through two essential layers: the frontend and the backend. These layers work together to transform human interaction into intelligent responses, allowing AI systems to function smoothly and effectively.

The frontend is the part of the system that users see and interact with. It includes screens, input tools, and the display of results. Whether a user is typing a question, speaking a command, or uploading an image, all interaction begins at the frontend.

Behind the scenes, the backend serves as the processing engine of the system. It receives input from the frontend, manages data, connects to the AI model, and performs the computations needed to generate a response. The backend is where the intelligence of the system operates, using models, algorithms, and infrastructure to interpret and respond to user requests.

This visual illustrates how these two layers work together as a complete system. It shows the flow of information from user input to AI processing and back to the user as a result. By understanding this structure, students can see that AI is not a single tool, but a coordinated system of layers working together.

In simple terms, the frontend is where interaction happens, and the backend is where intelligence is created.

Every AI system has two sides: what you see and what actually does the thinking.

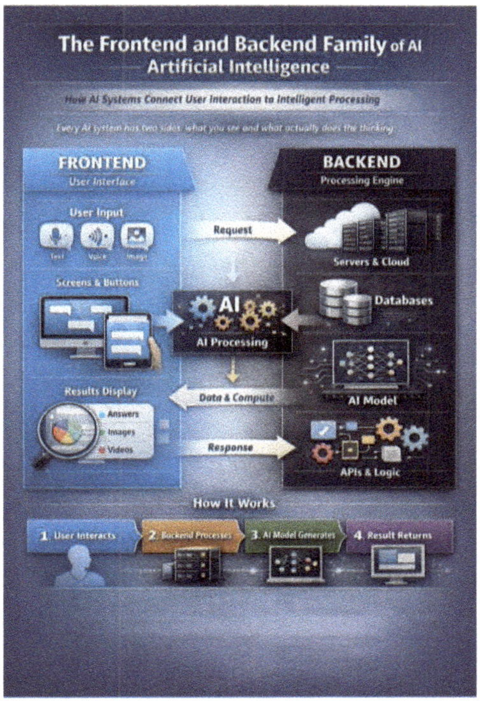

AI Qubit Family Introduction

The AI Qubit Family represents the foundation of quantum computing as it relates to artificial intelligence. While traditional computing relies on bits, which can exist as either 0 or 1, quantum computing uses qubits, which can exist in multiple states simultaneously.

A qubit is the basic unit of quantum information. Unlike classical bits, qubits can take advantage of properties such as superposition and entanglement, allowing quantum systems to process complex computations more efficiently in certain scenarios. This opens new possibilities for solving problems that are difficult or impossible for traditional computers.

In the context of artificial intelligence, qubits have the potential to accelerate learning processes, optimize large-scale systems, and improve complex simulations. Although quantum AI is still in development, it represents a future direction where computational power can expand beyond current limitations.

The AI Qubit Family connects with the Chip Family, as quantum processors are specialized hardware, and with the Algorithm Family, where new types of quantum algorithms are designed. It also relates to the Parameter and Neural Network Families, as researchers explore quantum-enhanced learning models.

Understanding the AI Qubit Family helps students see that artificial intelligence is continuously evolving. It introduces the idea that the foundations of computing themselves can change, leading to new forms of intelligence and problem-solving.

In simple terms, if bits are the foundation of today's computing, qubits represent the foundation of tomorrow's possibilities.

AI Qubit Family Breakdown Chart

Component	Role	Description	Example
Qubit	Quantum Unit	Basic unit of quantum information	Quantum bit in a quantum computer
Superposition	Multi-State Ability	Ability to exist in multiple states at once	0 and 1 simultaneously
Entanglement	Connection Property	Linking qubits so they influence each other	Paired quantum states
Quantum Gate	Operation System	Performs operations on qubits	Quantum logic gates
Quantum Circuit	Processing Structure	Sequence of quantum operations	Quantum algorithms
Quantum Processor	Hardware Engine	Specialized chip for quantum computing	Quantum computer hardware
Quantum Algorithm	Computation Method	Algorithm designed for quantum systems	Optimization problems
Quantum Speedup	Performance Advantage	Faster processing for certain problems	Complex simulations
Quantum Noise	Stability Challenge	Errors caused by environmental interference	Qubit instability
Quantum Error Correction	Stability Solution	Techniques to reduce errors in quantum systems	Fault-tolerant computing

The AI Qubit Family introduces a new frontier in computing, where the limits of traditional systems are expanded, opening the door to more powerful and advanced forms of artificial intelligence.

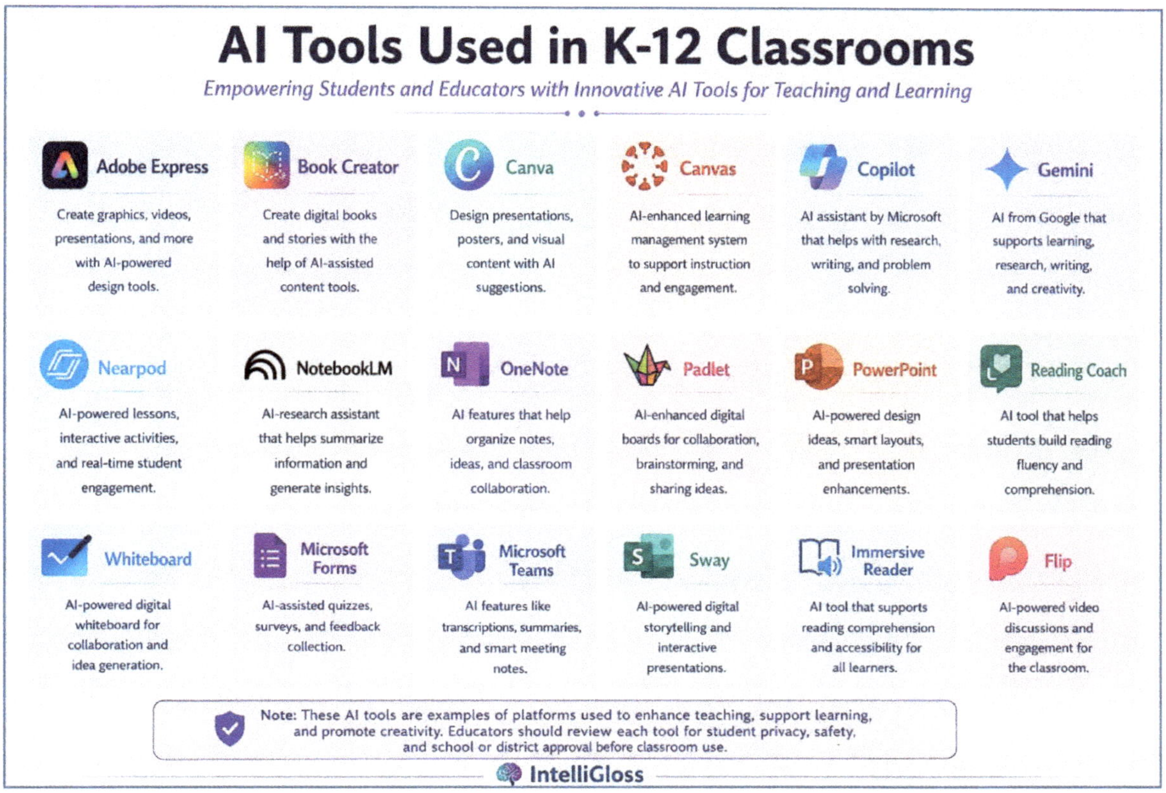
AI Tools Used in K-12 Classrooms
Empowering Students and Educators with Innovative AI Tools for Teaching and Learning
Adobe Express
Create graphics, videos, presentations, and more with AI-powered design tools.
Book Creator
Create digital books and stories with the help of AI-assisted content tools.
Canva
Design presentations, posters, and visual content with AI suggestions.
Canvas
AI-enhanced learning management system to support instruction and engagement.
Copilot
AI assistant by Microsoft that helps with research, writing, and problem solving.
Gemini
AI from Google that supports learning, research, writing, and creativity.
Nearpod
AI-powered lessons, interactive activities, and real-time student engagement.
NotebookLM
AI-research assistant that helps summarize information and generate insights.
OneNote
AI features that help organize notes, ideas, and classroom collaboration.
Padlet
AI-enhanced digital boards for collaboration, brainstorming, and sharing ideas.
PowerPoint
AI-powered design ideas, smart layouts, and presentation enhancements.
Reading Coach
AI tool that helps students build reading fluency and comprehension.
Whiteboard
AI-powered digital whiteboard for collaboration and idea generation.
Microsoft Forms
AI-assisted quizzes, surveys, and feedback collection.
Microsoft Teams
AI features like transcriptions, summaries, and smart meeting notes.
Sway
AI-powered digital storytelling and interactive presentations.
Immersive Reader
AI tool that supports reading comprehension and accessibility for all learners.
Flip
AI-powered video discussions and engagement for the classroom.
Note: These AI tools are examples of platforms used to enhance teaching, support learning, and promote creativity. Educators should review each tool for student privacy, safety, and school or district approval before classroom use.
IntelliGloss

TABLE OF CONTENTS

Introduction

Large Language Models: How Machines Learn Human Language

Human civilization has always advanced through language.
From ancient storytelling to scientific discoveries, language has been the bridge that carries knowledge from one mind to another and from one generation to the next. It allows us to teach, debate, imagine, and build societies based on shared understanding.

For centuries, language was considered one of the most uniquely human abilities.
While machines could calculate numbers and follow instructions, they could not truly understand the complexity of human communication—its grammar, context, tone, and meaning.

That is now changing.

In the early decades of the twenty-first century, a new class of artificial intelligence systems emerged known as **Large Language Models (LLMs)**. These systems are designed to analyze, interpret, and generate human language by learning patterns from vast collections of text.

Powered by advanced neural networks and enormous amounts of data, LLMs are capable of writing essays, answering questions, translating languages, generating computer code, assisting research, and helping people learn in entirely new ways.

Today, technologies built on large language models are becoming part of everyday life. Systems such as ChatGPT, Claude, and Gemini demonstrate how machines can participate in conversations, assist with problem-solving, and support learning across many fields.

But despite their growing presence, many people still ask an important question:

How do these systems actually work?

This book explores the science behind large language models in a clear and structured way designed for students, educators, and curious learners. Rather than focusing only on how to *use* AI tools, this book explains **how machines learn language in the first place**.

Inside these chapters, you will discover:

- How human language can be transformed into digital data
- Why tokens and patterns allow machines to interpret words
- How neural networks and transformer architectures power modern AI systems
- Why parameters allow models to store knowledge and generate responses
- What limitations and challenges still exist in current AI technologies
- How society must guide these systems responsibly and ethically

Understanding large language models is not just about technology—it is about understanding how knowledge itself is evolving.

For the first time in history, machines are able to assist humans in processing and generating vast amounts of language-based information. This creates new opportunities for education, research, creativity, and collaboration.

Yet with these opportunities also come responsibilities.
The future of intelligent systems will depend not only on engineers and scientists, but also on informed citizens who understand both the power and the limitations of artificial intelligence.

This book invites you to explore that future.

By learning how machines learn language, you will gain insight into one of the most important technological transformations of our time—and into the new partnership forming between **human intelligence and machine intelligence**.

Welcome to the world of **Large Language Models**.

Chapter 1 — The Power of Human Language

Learning Objectives

By the end of this chapter, students will be able to:

1. Understand why language is one of the most important abilities of human intelligence.
2. Explain how language allows humans to share knowledge, ideas, and culture.
3. Describe why teaching machines to understand language has been a major challenge in artificial intelligence.
4. Recognize how modern AI systems attempt to process human language.
5. Reflect on the role language plays in shaping learning, communication, and civilization.

1.1 Language: Humanity's Greatest Tool

Language is one of the most powerful inventions in human history.

Long before computers, satellites, or modern science, humans developed systems of language to communicate ideas, emotions, and experiences. Through language, people can share knowledge, tell stories, teach new skills, and record discoveries.

Language allows individuals to do something extraordinary:
transfer thoughts from one mind to another.

When a teacher explains a concept, a writer publishes a book, or a scientist shares research, language becomes the bridge that connects human intelligence across time and distance.

Every civilization in history has depended on language for survival and progress.

Through language, humans have been able to:

- Record history
- Develop laws and systems of government
- Build scientific knowledge
- Share cultural traditions
- Educate future generations

Without language, the accumulation of knowledge across generations would be impossible.

Language is therefore not just a communication tool—it is the **foundation of civilization itself.**

1.2 Language and the Human Mind

The human brain is uniquely designed to process language.

From an early age, children learn to recognize sounds, words, and sentence structures. Over time, they develop the ability to express complex thoughts and understand the ideas of others.

Language processing in the brain involves several cognitive abilities:

- **Memory**, which stores vocabulary and grammar
- **Pattern recognition**, which helps us understand sentence structure
- **Context awareness**, which allows us to interpret meaning
- **Emotion and empathy**, which help us understand tone and intention

For example, consider the sentence:

"It's cold in here."

Depending on the situation, this sentence could mean several things:

- A simple observation about temperature
- A request to close a window
- A signal that someone feels uncomfortable

Humans easily interpret these subtle meanings because our brains combine language with **context, experience, and social understanding.**

For machines, however, interpreting language has historically been extremely difficult.

1.3 The Challenge of Teaching Machines Language

Computers were originally designed to work with numbers and precise instructions.

A calculator, for example, performs arithmetic operations with perfect accuracy because numbers follow clear mathematical rules.

Language, however, is very different.

Human language contains:

- Ambiguity
- Multiple meanings for the same word
- Cultural references
- Emotional tone
- Complex grammar and structure

Consider the word **"bank."**

It could refer to:

- A financial institution
- The edge of a river
- A place where airplanes turn during flight

Humans easily understand which meaning applies based on context.
Machines, however, must learn to determine meaning by analyzing patterns in enormous amounts of data.

This challenge has been one of the central problems in artificial intelligence for decades.

Scientists have long asked:

Can a machine learn to understand human language?

1.4 The Evolution of Language Processing in AI

Early attempts at machine language processing relied on simple rule-based systems.

These systems followed strict grammatical rules programmed by engineers. While useful for basic tasks, they struggled to handle the complexity and unpredictability of real human communication.

As computing technology advanced, researchers began developing **machine learning systems** that could study large collections of text and identify patterns in language.

Instead of being told every rule, these systems could **learn from examples.**

This shift marked the beginning of a new era in artificial intelligence.

Modern AI language systems are built using advanced neural network architectures that analyze enormous datasets containing books, articles, conversations, and digital text from across the internet.

Through this process, machines gradually learn:

- How words relate to each other
- How sentences are structured
- How ideas flow within conversations

This technology eventually led to the development of **Large Language Models**, the subject of this book.

1.5 Large Language Models: A New Era of AI

Large Language Models, often called **LLMs**, represent one of the most significant breakthroughs in artificial intelligence.

These systems are trained on massive amounts of text and use deep neural networks to predict and generate language.

Instead of memorizing specific sentences, they learn statistical patterns that allow them to generate new responses.

This ability enables machines to perform tasks such as:

- Answering questions
- Writing essays
- Translating languages
- Generating computer code
- Assisting research and education

Many modern AI systems that people interact with today are powered by large language models, including systems such as ChatGPT, Claude, and Gemini.

While these systems can appear intelligent in conversation, it is important to understand that they do not think in the same way humans do.

Instead, they rely on mathematical patterns learned from enormous datasets.

Understanding how these systems work is essential for anyone living in the modern digital world.

1.6 Why Students Should Understand AI Language Systems

Artificial intelligence is becoming part of many areas of daily life.

Students today may encounter AI systems in:

- Education and tutoring platforms
- Research and writing assistance
- Translation tools
- Customer service systems
- Scientific research applications

Understanding how language models work allows students to use these tools responsibly and effectively.

More importantly, learning about AI helps students develop critical thinking skills about technology and its role in society.

The goal of this book is not only to explain how machines learn language, but also to encourage thoughtful reflection about how humans and intelligent systems will collaborate in the future.

The next chapter will explore how human language can be transformed into **data that machines can understand**, beginning the journey into the inner workings of large language models.

Chapter Summary

Language is one of the most powerful tools humans possess. It allows knowledge, ideas, and culture to be shared across generations and societies.

Teaching machines to understand human language has been a major challenge because language contains ambiguity, context, and emotional meaning.

Early computer systems struggled with language because they relied on rigid rules. Modern artificial intelligence systems instead learn from large collections of text, identifying patterns that allow them to generate and interpret language.

Large Language Models represent a major advancement in this field, enabling machines to assist humans in communication, research, and education.

Understanding these systems is an important step in preparing students for a future where human intelligence and machine intelligence work together.

The next chapter will explore how words and sentences can be converted into digital data that machines can analyze and learn from.

create a reflection section, worksheet, quiz and answer key

Partner, here are the **four educational sections for Chapter 1** formatted clearly for your school curriculum style.

Reflection Section

Chapter 1 – The Power of Human Language

Reflection questions encourage students to think deeply about the ideas discussed in the chapter and connect them to their own experiences.

1. Why is language considered one of the most powerful tools humans possess?
2. How does language help societies pass knowledge from one generation to the next?
3. Think about a time when communication helped you solve a problem or learn something new. What role did language play?
4. Why do you think it has been difficult for computers to understand human language?
5. How might artificial intelligence systems that understand language change education, communication, or work in the future?

Reflection Activity

Write a short paragraph describing how communication would change if humans suddenly lost the ability to use language. What challenges would society face?

Worksheet

Chapter 1 – The Power of Human Language

Part 1: Vocabulary Match

Match each term with the correct description.

Term	Description
Language	A. The ability of computers to perform tasks that normally require human intelligence
Artificial Intelligence	B. A system of communication using words and symbols
Context	C. The situation or background that helps determine meaning
Neural Network	D. A computer system designed to recognize patterns in data

Part 2: Short Answer

1. Explain why language is important for human civilization.
2. What makes human language difficult for machines to understand?
3. What are two ways large language models can assist people in everyday life?
4. How do machines learn patterns in language?
5. Why is it important for students to understand artificial intelligence systems?

Part 3: Critical Thinking

Imagine you are designing a robot that must communicate with humans.

Write 4–5 sentences describing what abilities the robot would need in order to understand human language effectively.

Quiz

Chapter 1 – The Power of Human Language

Multiple Choice

1. What is the primary purpose of language in human society?
   A. To perform calculations
   B. To store electricity
   C. To communicate ideas and knowledge
   D. To control machines

2. Why is language difficult for computers to understand?
   A. Language contains numbers only
   B. Language has context, emotion, and multiple meanings
   C. Computers cannot read text
   D. Language is always predictable

3. What is one goal of large language models?
   A. To build physical robots
   B. To generate and understand human language
   C. To manufacture computer chips
   D. To replace the internet

4. Large language models learn language by:
   A. Reading one book repeatedly
   B. Studying patterns in large collections of text
   C. Memorizing dictionaries
   D. Listening to music

5. Which of the following technologies uses large language models?
   A. Weather satellites
   B. Navigation systems
   C. AI chat assistants
   D. Video cameras

Short Answer

6. Describe one way language helps humans share knowledge.
7. Explain why context is important when interpreting words.

Quiz Answer Key

Multiple Choice

1. **C** — To communicate ideas and knowledge
2. **B** — Language has context, emotion, and multiple meanings
3. **B** — To generate and understand human language
4. **B** — Studying patterns in large collections of text
5. **C** — AI chat assistants

Short Answer (Example Answers)

6. Language allows people to teach, write books, share discoveries, and pass knowledge to future generations.
7. Context helps determine the correct meaning of words that may have multiple interpretations depending on the situation.

Chapter 2 — From Words to Data: Turning Language into Information Machines Can Understand

Learning Objectives

By the end of this chapter, students will be able to:

1. Explain how human language can be converted into digital data.
2. Describe how computers represent words and sentences numerically.
3. Understand why machines must transform language into numbers before processing it.

4. Recognize the importance of datasets in training artificial intelligence systems.
5. Explain how data enables large language models to learn patterns in human communication.

2.1 The Digital Nature of Computers

Computers do not understand language the way humans do.

Humans read words, hear tone, and interpret meaning through experience and emotion. Machines, however, operate through **mathematical instructions and digital signals**.

At the most basic level, computers process information using **binary code**, a system made of two symbols: **0 and 1**.

Every image, video, document, or message stored on a computer is ultimately represented as combinations of these binary signals.

This means that before a machine can work with language, something important must happen:

Words must first be converted into digital data.

This transformation allows machines to analyze, compare, and learn patterns from text.

Without converting language into data, artificial intelligence systems would have no way to process written or spoken communication.

2.2 Transforming Language into Machine-Readable Form

When humans read a sentence, we interpret meaning instantly.

For example:

"The student opened the book."

A person easily understands the action, the subject, and the object in the sentence.

For a computer, however, this sentence initially appears as a string of characters.

To make sense of it, computers must break the sentence into parts and convert each part into a numerical form that algorithms can process.

This process involves several steps:

1. **Text Input** – A sentence or document is provided to the system.
2. **Text Processing** – The system analyzes the structure of the sentence.
3. **Numerical Encoding** – Words are converted into numbers or vectors.

4. **Pattern Analysis** – The machine studies relationships between words.

By turning language into numbers, machines can perform mathematical operations that reveal patterns in communication.

2.3 Why Machines Need Numbers

Machines rely on mathematics to perform every operation.

Unlike humans, computers cannot interpret meaning directly from letters or sounds. Instead, they rely on **numerical representations of information**.

For example, a simple system might assign numbers to words such as:

- "cat" → 125
- "dog" → 209
- "tree" → 512

This numeric representation allows machines to store and compare words.

However, modern artificial intelligence systems go far beyond simple numbering.

Large language models represent words using **vectors**, which are sets of numbers that capture relationships between words.

For example, words that appear in similar contexts—such as "king" and "queen"—may have similar vector representations.

This allows AI systems to detect meaning through mathematical relationships.

In this way, numbers become the **language that machines use to understand human communication.**

2.4 The Role of Datasets in Language Learning

Large language models learn by analyzing enormous collections of text called **datasets**.

A dataset is a structured collection of information used to train machine learning systems.

For language models, datasets often contain:

- Books
- Articles
- Websites
- Academic papers
- Conversations and written dialogue

By studying these texts, AI systems observe how words appear together, how sentences are structured, and how ideas flow within language.

The more data a model studies, the better it becomes at predicting how language works.

This process is similar to how humans learn language.

Children learn words and grammar by hearing and reading thousands of examples over time. In a similar way, machines analyze large volumes of text to learn patterns in communication.

2.5 Patterns in Language Data

Language may seem unpredictable, but it actually contains many patterns.

For example, certain words frequently appear together:

- "peanut butter and jelly"
- "cause and effect"
- "research and development"

Machines learn these patterns by examining billions of sentences.

Through statistical analysis, language models identify relationships such as:

- Word order in sentences
- Common phrases
- Grammatical structures
- Topic associations

These patterns allow AI systems to predict what words are likely to appear next in a sentence.

For example, if a sentence begins with:

"The sun rises in the..."

A language model might predict that the next word will be **"east."**

This prediction process is the foundation of how large language models generate responses.

2.6 From Data to Understanding

Although language models analyze data mathematically, their outputs can appear remarkably human-like.

When trained on vast datasets, modern AI systems can produce essays, answer questions, summarize information, and assist with learning.

Some well-known AI systems that rely on large language models include
ChatGPT,
Claude, and
Gemini.

These systems analyze user input, convert it into numerical form, and generate responses by predicting patterns in language.

However, it is important to remember that these systems do not truly **understand language in the human sense**.

Instead, they rely on statistical relationships learned from data.

Understanding this distinction helps students use AI tools responsibly and critically.

Chapter Summary

Computers cannot directly interpret human language.

Before machines can process words and sentences, language must first be converted into digital data that computers can analyze mathematically.

This process involves transforming text into numerical representations that allow algorithms to detect patterns in communication.

Large language models rely on enormous datasets to learn how words and ideas are connected.

By studying billions of examples, these systems identify patterns in grammar, vocabulary, and sentence structure.

Although AI systems can generate impressive language-based responses, they operate through mathematical predictions rather than human-like understanding.

The next chapter will explore the concept of **tokens**, the fundamental units that allow machines to break language into manageable pieces for analysis and learning.

can you create a reflection section, worksheet, quiz and answer key

Reflection Section

Chapter 2 – From Words to Data: Turning Language into Information Machines Can Understand

Reflection helps students think about how the ideas in this chapter apply to the real world.

1. Why must computers convert human language into numbers before processing it?

2. How is the way machines learn language similar to how humans learn language?
3. Why are large datasets important when training artificial intelligence systems?
4. What kinds of information might appear in a dataset used to train language models?
5. Do you think machines truly understand language, or do they simply recognize patterns? Explain your reasoning.

Reflection Activity

Imagine a future where machines can instantly translate any language in the world.
Write a short paragraph explaining how this technology might change global communication, education, and international collaboration.

Worksheet

Chapter 2 – From Words to Data

Part 1: Vocabulary Match

Match each term with the correct definition.

Term	Definition
Dataset	A. A numerical representation of words used by AI systems
Binary Code	B. A collection of information used to train machine learning systems
Vector	C. A digital system using 0 and 1 to represent information
Encoding	D. The process of converting text into a format computers can process

Part 2: Fill in the Blank

1. Computers process information using a digital system called __________.
2. Before machines can analyze language, words must be converted into __________.
3. A large collection of information used to train AI is called a __________.
4. AI systems learn language by identifying __________ in text data.
5. Words in AI systems are often represented as sets of numbers called __________.

Part 3: Short Answer

1. Explain why computers rely on numbers instead of words to process information.
2. What types of materials might be included in datasets used to train language models?
3. Why do larger datasets usually help improve AI performance?
4. What is one similarity between how humans learn language and how AI models learn language?

Part 4: Critical Thinking

Imagine you are responsible for building a dataset to train a language model for students.

Write 4–5 sentences explaining what types of information you would include and why.

Quiz

Chapter 2 – From Words to Data

Multiple Choice

1. Why must computers convert language into numbers?
   A. Because computers cannot store letters
   B. Because computers operate using mathematical calculations
   C. Because words are too long
   D. Because language cannot be stored digitally

2. What is a dataset?
   A. A group of computer chips
   B. A large collection of information used to train AI systems
   C. A type of programming language
   D. A storage device

3. What digital system do computers use to process information?
   A. Alphabetic code
   B. Binary code
   C. Morse code
   D. Symbolic code

4. Why do AI systems analyze large amounts of text?
   A. To memorize books
   B. To learn patterns in language
   C. To store every sentence
   D. To translate images

5. What mathematical structure often represents words in AI models?
   A. Images
   B. Vectors
   C. Charts
   D. Graphs

Short Answer

6. What is one example of information that could be included in an AI dataset?
7. Why is recognizing patterns important for machine learning?

Quiz Answer Key

Multiple Choice

1. **B** — Because computers operate using mathematical calculations
2. **B** — A large collection of information used to train AI systems
3. **B** — Binary code
4. **B** — To learn patterns in language
5. **B** — Vectors

Fill in the Blank (Worksheet)

1. **Binary code**
2. **Numbers**
3. **Dataset**
4. **Patterns**
5. **Vectors**

Short Answer (Example Responses)

6. Examples may include books, articles, websites, academic papers, or conversations.
7. Recognizing patterns allows AI systems to predict how language works and generate meaningful responses.

Chapter 3 — Tokens: The Building Blocks of AI Language

Learning Objectives

By the end of this chapter, students will be able to:

1. Define what tokens are in the context of artificial intelligence and language models.
2. Explain why AI systems break text into smaller units called tokens.
3. Understand how tokenization allows machines to process human language efficiently.
4. Recognize the role tokens play in training large language models.
5. Describe how token-based processing helps machines analyze patterns in communication.

3.1 Breaking Language into Manageable Pieces

Human language is incredibly complex. Sentences can contain dozens of words, and entire documents may contain thousands or even millions of characters.

While humans can easily read long passages of text, computers must process information in a structured and systematic way. For this reason, artificial intelligence systems divide language into smaller units that are easier to analyze.

These units are called **tokens**.

A token is a small piece of language that an AI system uses as a basic unit of analysis. Tokens allow machines to break large amounts of text into manageable parts that can be processed mathematically.

Tokens may represent:

- Whole words
- Parts of words
- Individual characters
- Punctuation marks
- Numbers or symbols

By dividing language into tokens, machines can analyze the structure of sentences and identify patterns that help them understand how language works.

This process is one of the most important steps in building modern language models.

3.2 What Is Tokenization?

The process of converting text into tokens is called **tokenization**.

Tokenization is the step where a computer takes a sentence and separates it into individual language units that can be analyzed by machine learning algorithms.

For example, consider the sentence:

"Artificial intelligence is transforming education."

A tokenized version of this sentence might appear as:

Artificial | intelligence | is | transforming | education | .

Each word becomes a token, and punctuation marks may also be treated as tokens.

However, tokenization is not always this simple.

In many modern language models, tokens may represent **parts of words** rather than entire words. This approach helps machines handle large vocabularies and unfamiliar terms more effectively.

For example, the word **"transforming"** might be split into tokens such as:

transform | ing

This method allows language models to recognize shared patterns between related words such as:

- transform
- transformation
- transforming
- transformed

Tokenization therefore helps machines understand language in a flexible and efficient way.

3.3 Why Tokens Are Necessary for AI

Artificial intelligence systems cannot process entire documents as a single unit.

Instead, they must examine language step by step.

Tokens provide the structure that allows machines to analyze text systematically.

Using tokens, AI systems can:

- Track the position of words within sentences

- Identify relationships between words
- Analyze grammar and syntax
- Predict the next word in a sequence

For example, if a language model receives the phrase:

"The capital of France is..."

The system analyzes the tokens in the sentence and predicts the most likely next token based on patterns learned during training.

Because the model has observed many examples in its training data, it may predict that the next token will be **"Paris."**

In this way, tokens enable language models to generate meaningful responses through probability and pattern recognition.

3.4 Tokens and Vocabulary in Language Models

Every language model operates with a predefined collection of tokens called a **token vocabulary**.

This vocabulary contains thousands or sometimes hundreds of thousands of tokens that represent the building blocks of language used by the model.

The vocabulary typically includes:

- Common words
- Word fragments
- Numbers
- Punctuation marks
- Special symbols

When a user enters text into an AI system, the model converts the input into tokens from this vocabulary.

If a word does not exist in the vocabulary, the system breaks the word into smaller token components that it recognizes.

For example, a rare word such as **"microarchitecture"** might be divided into tokens such as:

micro | architecture

This approach allows language models to process unfamiliar words while still recognizing meaningful patterns.

3.5 Tokens and Context

Tokens do not exist in isolation. Their meaning depends heavily on the surrounding tokens in a sentence.

For example, consider the word **“light.”**

The meaning changes depending on the sentence:

- "The room needs more light."
- "The backpack feels light."

Although the same token appears in both sentences, the surrounding tokens help determine its meaning.

Modern language models analyze tokens in relation to the tokens that come before and after them.

By studying these relationships, AI systems learn how meaning changes depending on context.

This contextual understanding is one of the key reasons modern language models are able to produce more coherent and natural responses.

3.6 Tokens in Modern Language Models

Tokens are the fundamental units that allow large language models to process and generate language.

When a user types a prompt into an AI system, the following process occurs:

1. The input text is converted into tokens.
2. Each token is transformed into numerical data.
3. The AI system analyzes relationships between tokens.
4. The system predicts the most likely next token.
5. The predicted tokens are converted back into readable language.

This cycle continues repeatedly until the system produces a complete response.

Many modern AI tools that people interact with daily rely on token-based language processing systems, including systems such as ChatGPT, Claude, and Gemini.

Although these systems can appear conversational and intelligent, their responses are generated through mathematical predictions based on token relationships learned during training.

Understanding tokens therefore provides a foundational insight into how language models function.

Chapter Summary

Tokens are the basic units that allow artificial intelligence systems to process human language.

Because computers cannot analyze entire sentences or documents as a single piece of information, language must first be divided into smaller components.

The process of dividing text into tokens is called tokenization.

Tokens may represent words, parts of words, punctuation, or symbols. These units allow machines to analyze the structure of language and identify patterns within large datasets.

Language models rely on token vocabularies to convert human text into numerical data that algorithms can analyze. By studying relationships between tokens, AI systems learn how language works and generate responses based on probability and context.

Tokens therefore serve as the **foundation of modern language processing in artificial intelligence**.

In the next chapter, we will explore how machines use tokens to detect **patterns in language**, allowing AI systems to predict meaning and generate coherent responses.

Reflection Section

Chapter 3 – Tokens: The Building Blocks of AI Language

Reflection questions help students connect the chapter concepts to their own understanding of language and technology.

1. Why do artificial intelligence systems break language into tokens instead of analyzing full sentences at once?
2. How does tokenization help computers manage large amounts of text data?
3. Why might it be useful for a language model to break words into smaller parts rather than treating every word as a separate unit?
4. Think about the way humans read sentences. How is human reading different from how machines process tokens?
5. How might token-based processing help AI systems generate responses in conversations?

Reflection Activity

Write a short paragraph explaining how breaking information into smaller pieces can help solve complex problems. You may compare this idea to learning a new subject, solving a puzzle, or organizing a large project.

Worksheet

Chapter 3 – Tokens: The Building Blocks of AI Language

Part 1: Vocabulary Match

Match each term with the correct definition.

Term	Definition
Token	A. The collection of language units recognized by a language model
Tokenization	B. A small piece of language used by AI systems for processing text
Vocabulary	C. The process of breaking text into smaller units
Context	D. The surrounding words that influence meaning

Part 2: Fill in the Blank

1. A __________ is a small unit of language used by AI systems to analyze text.
2. The process of breaking text into tokens is called __________.
3. Language models analyze relationships between __________ to understand language patterns.
4. The meaning of a token often depends on its __________ in a sentence.
5. A model's token __________ contains the language units it recognizes.

Part 3: Short Answer

1. Explain why tokenization is necessary for machines to process language.
2. What types of elements can tokens represent in a sentence?
3. Why might large language models divide unfamiliar words into smaller tokens?
4. How does tokenization help AI systems identify patterns in language?

Part 4: Critical Thinking

Imagine you are designing an AI system that reads stories and answers questions.

Write 4–5 sentences explaining how tokenization would help the system understand and analyze the text.

Quiz

Chapter 3 – Tokens: The Building Blocks of AI Language

Multiple Choice

1. What is a token in artificial intelligence language processing?
   A. A computer chip
   B. A small unit of language used by AI systems
   C. A type of algorithm
   D. A programming command

   What is tokenization?
   A. Converting images into text
   B. Breaking text into smaller language units
   C. Storing language in a database
   D. Translating language automatically

3. Why do language models use tokens?
   A. To organize text into manageable units for analysis
   B. To store electricity
   C. To reduce computer memory
   D. To display graphics

4. What can tokens represent in text?
   A. Words only
   B. Characters only
   C. Words, parts of words, punctuation, or symbols
   D. Images

5. Why is context important in language processing?
   A. It changes the meaning of words depending on surrounding language
   B. It stores data in memory
   C. It reduces vocabulary size
   D. It speeds up computers

Short Answer

6. What is one advantage of breaking language into tokens?
7. Why might a language model break long or unfamiliar words into smaller tokens?

Quiz Answer Key

Multiple Choice

1. **B** — A small unit of language used by AI systems
2. **B** — Breaking text into smaller language units
3. **A** — To organize text into manageable units for analysis
4. **C** — Words, parts of words, punctuation, or symbols
5. **A** — It changes the meaning of words depending on surrounding language

Fill in the Blank (Worksheet)

1. **Token**
2. **Tokenization**
3. **Tokens**
4. **Context**
5. **Vocabulary**

Short Answer (Example Responses)

6. Breaking language into tokens allows AI systems to analyze text step by step and identify patterns more easily.
7. Dividing unfamiliar words into smaller tokens allows the model to recognize parts of the word that appear in other contexts, helping it understand meaning more effectively.

Chapter 4 — Patterns in Language: How Machines Learn Meaning

Learning Objectives

By the end of this chapter, students will be able to:

1. Explain why patterns are essential for understanding language.
2. Describe how artificial intelligence systems detect patterns in text data.

3. Understand how pattern recognition helps machines predict words and generate responses.
4. Recognize the difference between memorizing language and learning patterns in language.
5. Explain why pattern recognition is one of the most important skills in artificial intelligence.

4.1 Language Is Built on Patterns

Human language may appear spontaneous and creative, but it is built upon many underlying patterns.

Every language follows rules of grammar, sentence structure, and word relationships. Even when people speak casually, these patterns guide how sentences are formed and understood.

For example, consider the sentence:

"The scientist conducted an experiment."

This sentence follows a common pattern in English:

Subject → Verb → Object

- Subject: the scientist
- Verb: conducted
- Object: an experiment

Most sentences in English follow similar grammatical structures. Because humans learn these patterns from childhood, we rarely think about them consciously.

However, these patterns are extremely important when teaching machines to process language.

Artificial intelligence systems must learn to recognize these patterns in order to understand how words and ideas are connected.

4.2 Pattern Recognition in Artificial Intelligence

Pattern recognition is one of the central concepts in machine learning and artificial intelligence.

Rather than memorizing every sentence ever written, AI systems analyze large amounts of text to identify recurring structures in language.

For example, after studying millions of sentences, a language model may observe that certain words frequently appear together, such as:

- cause and effect
- research and development
- law and order
- supply and demand

These recurring combinations help machines learn how language is typically used.

In addition to word pairings, AI systems also detect patterns in:

- sentence structure
- grammar
- topic relationships
- conversational flow

By analyzing these patterns, machines gradually learn how language behaves.

This process allows AI systems to generate responses that appear natural and coherent.

4.3 Learning Through Probability

Modern language models rely heavily on probability when generating text.

Probability refers to the likelihood that a particular word or phrase will appear next in a sequence.

For example, consider the sentence:

"Water freezes at zero degrees..."

Based on patterns in language data, a model might predict that the next word is likely to be **"Celsius."**

This prediction does not come from memorizing the sentence directly. Instead, it results from observing similar patterns across large datasets.

The model calculates probabilities for many possible next words and selects the most likely option.

This process occurs repeatedly as the model generates each word in a response.

As a result, language models produce sentences by building them **one token at a time**, guided by statistical patterns.

4.4 Context and Meaning

Patterns in language are not determined by individual words alone. The meaning of a sentence depends heavily on context.

Context refers to the surrounding words, phrases, and ideas that help determine meaning.

Consider the word **"bat."**

It could refer to:

- a flying mammal
- a piece of sports equipment used in baseball

Humans determine the correct meaning based on the context of the sentence.

For example:

- *"The bat flew across the night sky."*
- *"The player swung the bat."*

Artificial intelligence systems must also analyze context in order to interpret meaning correctly.

Language models do this by examining the relationships between tokens within a sentence.

By analyzing surrounding tokens, AI systems determine which interpretation is most likely in a given situation.

4.5 Learning Patterns from Large Datasets

Large language models learn patterns by studying enormous collections of text.

These collections, known as **training datasets**, may include:

- books
- scientific articles
- news reports
- online discussions
- educational materials

By analyzing billions of examples, AI systems begin to recognize:

- how sentences are constructed
- how ideas are connected
- how topics develop across paragraphs

The scale of this data allows models to learn extremely complex language patterns.

However, it is important to understand that language models do not truly understand the meaning of information in the same way humans do.

Instead, they detect statistical relationships between words and phrases.

These relationships allow the system to generate responses that appear meaningful.

4.6 Pattern Recognition in Modern AI Systems

Many modern artificial intelligence tools rely on pattern recognition to process language.

Systems such as ChatGPT, Claude, and Gemini analyze patterns in user input and generate responses based on learned relationships within language data.

When a user asks a question, the system:

1. Converts the input into tokens.
2. Analyzes patterns in the tokens.
3. Calculates probabilities for possible responses.
4. Generates an answer based on the most likely sequence of tokens.

This process occurs extremely quickly, allowing AI systems to respond almost instantly.

Although these systems can produce impressive results, they still rely entirely on patterns learned during training.

Understanding this limitation helps users interpret AI-generated responses critically and responsibly.

Chapter Summary

Human language is governed by many patterns in grammar, vocabulary, and sentence structure.

Artificial intelligence systems rely on pattern recognition to analyze language and generate responses.

Rather than memorizing individual sentences, language models study enormous datasets to detect recurring structures and relationships between words.

Using probability and contextual analysis, these systems predict which tokens are most likely to appear next in a sequence.

This ability allows large language models to generate coherent and useful responses across many different tasks.

However, it is important to remember that AI systems operate through statistical pattern recognition rather than true human understanding.

In the next chapter, we will explore the **history of language models**, examining how early systems evolved into the powerful large language models used today.

create a reflection section, worksheet, quiz and answer key

Reflection Section

Chapter 4 – Patterns in Language: How Machines Learn Meaning

Reflection encourages students to connect the ideas in the chapter to their own experiences with language and learning.

1. Why are patterns important in human language?
2. How do artificial intelligence systems use patterns to generate responses?
3. Think about how children learn language. How is this similar to how machines learn patterns in language?
4. Why is context important when interpreting the meaning of a word?
5. Why is it important for students to understand that AI systems rely on probability rather than true understanding?

Reflection Activity

Write a short paragraph explaining how recognizing patterns helps humans learn new skills, such as reading, mathematics, or music. Then explain how similar pattern recognition helps artificial intelligence systems process language.

Worksheet

Chapter 4 – Patterns in Language

Part 1: Vocabulary Match

Match each term with the correct definition.

Term	Definition
Pattern Recognition	A. The surrounding words that influence meaning
Probability	B. The likelihood that something will occur
Context	C. The ability of a system to identify repeated structures in data
Dataset	D. A large collection of information used to train AI systems

Part 2: Fill in the Blank

1. __________ refers to identifying repeated structures or relationships in data.
2. AI systems study large collections of text called __________ to learn language patterns.
3. The meaning of a word often depends on its __________ in a sentence.
4. Language models generate responses by predicting the most likely __________ in a sequence.
5. __________ refers to the likelihood that a certain event will occur.

Part 3: Short Answer

1. Explain why recognizing patterns is important for understanding language.
2. How do large language models learn patterns from data?
3. What role does probability play in language generation?
4. Why is context important when determining the meaning of a word?

Part 4: Critical Thinking

Imagine you are teaching a computer how to complete the sentence:

"The Earth revolves around the ____."

Write 4–5 sentences explaining how pattern recognition and probability could help the computer predict the correct answer.

Quiz

Chapter 4 – Patterns in Language

Multiple Choice

1. What is pattern recognition?
   A. Storing large amounts of information
   B. Identifying repeated structures or relationships in data
   C. Translating languages automatically
   D. Designing computer hardware

2. Why do AI systems analyze large datasets of text?
   A. To memorize books
   B. To detect patterns in language
   C. To store sentences permanently
   D. To generate images

3. What does probability represent in language models?
   A. The speed of a computer
   B. The likelihood that a word or phrase will appear next
   C. The size of a dataset
   D. The number of tokens in a sentence

4. Why is context important in language processing?
   A. It changes the meaning of words depending on surrounding language
   B. It stores text in memory
   C. It reduces the number of words in a sentence
   D. It speeds up computer processing

5. What do modern language models rely on to generate responses?
   A. Pattern recognition and probability
   B. Manual programming only
   C. Memorized dictionaries
   D. Random word selection

Short Answer

6. Give one example of a common language pattern.
7. Why is it important to understand the limitations of AI pattern recognition?

Quiz Answer Key

Multiple Choice

1. **B** — Identifying repeated structures or relationships in data
2. **B** — To detect patterns in language
3. **B** — The likelihood that a word or phrase will appear next

4. **A** — It changes the meaning of words depending on surrounding language
5. **A** — Pattern recognition and probability

Fill in the Blank (Worksheet)

1. **Pattern recognition**
2. **Datasets**
3. **Context**
4. **Token**
5. **Probability**

Short Answer (Example Responses)

6. Examples may include phrases such as "cause and effect," "research and development," or sentence structures like subject–verb–object.
7. Understanding AI limitations helps people evaluate AI responses critically and recognize that AI relies on statistical patterns rather than true human understanding.

Chapter 5 — The Rise of Language Models: The Evolution of Machines That Understand Language

Learning Objectives

By the end of this chapter, students will be able to:

1. Explain how early computer systems attempted to process human language.
2. Describe how language models evolved from simple rule-based systems to modern machine learning systems.
3. Understand the role of data and computing power in advancing language technologies.
4. Recognize the technological milestones that led to the development of large language models.
5. Explain why modern AI systems are far more capable than earlier language-processing systems.

5.1 Early Attempts at Machine Language

For much of the twentieth century, computers were designed primarily for mathematical calculations and data processing. Early computers could perform arithmetic operations with remarkable speed, but they had little ability to understand human language. Engineers and scientists soon realized that if machines could interpret written or spoken language, computers could become far more useful in education, research, and communication.

The first efforts to teach computers language relied on what are known as **rule-based systems**. In these systems, programmers manually created large sets of linguistic rules that told the computer how to interpret sentences. These rules described grammatical structures such as nouns, verbs, and sentence order.

For example, a rule-based system might be programmed to recognize that a sentence often contains a subject followed by a verb and an object. When a user entered a sentence, the computer would attempt to match the sentence to its predefined grammatical rules.

Although these systems represented an important first step, they had serious limitations. Human language contains countless variations, exceptions, and creative expressions. As a result, rule-based systems often struggled when encountering unfamiliar phrases or complex sentences. Because every rule had to be programmed manually, expanding these systems became increasingly difficult.

These challenges led researchers to search for new methods that would allow machines to learn language more flexibly.

5.2 The Shift Toward Statistical Language Models

During the late twentieth century, researchers began exploring a different approach to language processing. Instead of programming every grammatical rule by hand, scientists began developing **statistical language models** that could learn patterns directly from data.

Statistical models analyze large collections of text and calculate how frequently words appear together. By studying these patterns, the system can estimate the probability that a certain word will appear after another word.

For example, after analyzing many sentences, a statistical language model might learn that the phrase *"New York"* appears frequently together. If the model encounters the word *"New,"* it may assign a high probability that the next word will be *"York."*

This method allows computers to generate predictions based on patterns observed in large datasets rather than relying solely on manually written rules.

Statistical language models marked an important turning point in natural language processing because they allowed machines to learn from real examples of human communication.

5.3 The Growth of Digital Data

The rise of the internet dramatically increased the amount of digital text available for analysis. Online articles, books, forums, and social media created enormous collections of written language that researchers could use to train machine learning systems.

This rapid expansion of digital information made it possible for language models to study patterns in millions and eventually billions of sentences. As datasets grew larger, language models became increasingly accurate in predicting word sequences and generating text.

At the same time, advances in computer hardware allowed machines to process larger datasets more efficiently. Faster processors and specialized computing devices made it possible to train more complex machine learning models.

The combination of large datasets and powerful computing resources created the foundation for the next generation of language technologies.

5.4 Neural Networks and Language Learning

A major breakthrough in artificial intelligence occurred with the development of **neural networks**. Neural networks are computational systems inspired by the structure of the human brain. Instead of relying on fixed rules, neural networks learn by adjusting internal numerical parameters based on training data.

In language processing tasks, neural networks analyze relationships between words across large datasets. Through repeated training, the system gradually improves its ability to predict words and understand sentence structure.

Neural networks proved especially effective at recognizing complex patterns that earlier statistical models could not easily capture. For example, neural networks can learn relationships between words that appear far apart in a sentence, allowing them to interpret context more effectively.

These capabilities significantly improved machine translation, text summarization, and conversational AI systems.

5.5 The Emergence of Large Language Models

As computing power continued to increase, researchers began building larger and more sophisticated neural network models. These systems were trained on extremely large datasets and contained billions of adjustable parameters that allowed them to learn detailed patterns in language.

These advanced systems became known as **large language models**, often abbreviated as LLMs.

Large language models are designed to process and generate language by analyzing patterns across massive collections of text. Because of their size and training scale, they can perform a wide range of language-related tasks, including answering questions, generating written content, summarizing information, and assisting with research.

Modern AI systems based on large language models are now widely used across education, science, business, and technology. Systems such as ChatGPT, Claude, and Gemini demonstrate how these technologies can interact with users in natural language.

While these systems may appear highly intelligent, it is important to remember that they still rely on statistical patterns learned from data rather than human-like understanding.

5.6 A New Era of Human–Machine Communication

The development of large language models marks an important milestone in the history of computing. For the first time, machines can participate in extended conversations, assist with writing tasks, and provide explanations across a wide range of topics.

These capabilities have opened new possibilities for education, research, and global communication. Students can now interact with AI systems to explore new ideas, ask questions, and receive guidance in their learning.

At the same time, the growing influence of language-based AI technologies raises important questions about responsible use, accuracy, and the role of human oversight.

Understanding the historical development of language models helps students recognize that these systems did not appear suddenly. They are the result of decades of scientific research, technological innovation, and the accumulation of digital knowledge.

Chapter Summary

Chapter 5 – The Rise of Language Models: The Evolution of Machines That Understand Language

The development of machines capable of processing human language has taken many decades of research and technological advancement. Early computer systems were designed primarily to perform mathematical calculations and follow strict instructions. Because of this limitation, early attempts at machine language processing relied on rule-based systems in which programmers manually wrote grammatical rules that computers could follow.

Although these systems represented an important early step in language technology, they struggled with the complexity and flexibility of human language. Human communication contains many variations, idioms, and contextual meanings that rule-based systems could not easily handle.

To overcome these limitations, researchers began developing statistical language models. These models analyze large collections of text and calculate the probability that certain words will appear together in sentences. By learning from patterns found in data, statistical models allowed computers to make predictions about language rather than relying solely on preprogrammed rules.

The rapid growth of digital information on the internet created massive datasets that could be used to train language models. At the same time, improvements in computing power made it possible to process larger amounts of data and build more advanced machine learning systems.

The introduction of neural networks marked a major breakthrough in artificial intelligence. Neural networks allowed computers to learn complex patterns in language by adjusting internal parameters during training. These systems could capture deeper relationships between words and sentences, leading to improved performance in tasks such as translation, summarization, and question answering.

These technological advances eventually led to the creation of large language models. Large language models are trained on enormous datasets and contain billions of adjustable parameters that allow them to generate and interpret human language with remarkable flexibility.

Today, language-based AI systems are used in many areas of society, including education, research, and business. Understanding the history of language models helps students appreciate how these technologies developed and why they continue to evolve.

Reflection Section

Chapter 5 – The Rise of Language Models

1. Why were early rule-based language systems limited in their ability to understand human language?
2. How did statistical language models improve the ability of machines to process text?
3. Why did the growth of digital data on the internet play an important role in the development of language models?
4. How do neural networks help machines learn patterns in language?
5. Why is it important for students to understand the historical development of artificial intelligence technologies?

Reflection Activity

Write a short paragraph explaining why technological advancements often occur gradually over time rather than appearing suddenly. Use the development of language models as an example in your explanation.

Worksheet

Chapter 5 – The Rise of Language Models

Part 1: Vocabulary Match

Match each term with the correct definition.

Term	Definition
Rule-Based System	A. A machine learning system inspired by the structure of the human brain
Statistical Language Model	B. A system that follows predefined grammatical rules
Neural Network	C. A system that predicts word relationships based on probability
Dataset	D. A large collection of information used for training AI systems

Part 2: Fill in the Blank

1. Early language-processing systems relied on __________ that were programmed by engineers.
2. Statistical language models analyze __________ in large collections of text.
3. A __________ is a collection of information used to train machine learning systems.
4. __________ allow machines to learn patterns in data by adjusting internal parameters.
5. Modern AI systems known as __________ are trained on massive datasets to generate language.

Part 3: Short Answer

1. Explain why rule-based systems struggled with human language.
2. What advantage did statistical language models provide compared to earlier rule-based systems?
3. Why are large datasets important for training language models?
4. How do neural networks improve language processing?

Part 4: Critical Thinking

Imagine you are part of a team designing a new language-processing system for students.

Write 4–5 sentences explaining what features or capabilities you would want the system to have in order to help students learn effectively.

Quiz

Chapter 5 – The Rise of Language Models

Multiple Choice

1. What type of system relied on manually programmed language rules?
   A. Neural network systems
   B. Rule-based systems
   C. Statistical systems
   D. Token systems

2. What do statistical language models analyze in order to predict words?
   A. Images
   B. Word patterns in text
   C. Computer hardware
   D. Electrical signals

3. What technological development helped language models learn complex relationships between words?
   A. Binary code
   B. Neural networks
   C. Computer keyboards
   D. Programming languages

4. What is a dataset used for in artificial intelligence?
   A. To store computer hardware
   B. To train machine learning systems
   C. To design computer chips
   D. To program operating systems

5. What are large language models designed to do?
   A. Generate and process human language
   B. Manufacture computer hardware
   C. Store electricity
   D. Build computer networks

Short Answer

6. Why did the expansion of digital data help advance language models?
7. Why are neural networks considered an important advancement in artificial intelligence?

Answer Key

Multiple Choice

1. **B** — Rule-based systems
2. **B** — Word patterns in text
3. **B** — Neural networks
4. **B** — To train machine learning systems
5. **A** — Generate and process human language

Fill in the Blank

1. **rules**
2. **patterns**
3. **dataset**
4. **Neural networks**
5. **large language models**

Short Answer (Example Responses)

6. The expansion of digital data provided large amounts of text that AI systems could analyze to learn patterns in language.
7. Neural networks allow machines to learn complex patterns in data by adjusting internal parameters during training, which improves their ability to process language and other information.

Chapter 6 — Neural Networks and Language Learning: How Machines Learn from Data

Learning Objectives

By the end of this chapter, students will be able to:

1. Explain what neural networks are and how they relate to artificial intelligence.
2. Describe how neural networks are inspired by the structure of the human brain.
3. Understand how neural networks learn patterns from large datasets.
4. Explain the role of layers and connections in neural network learning.
5. Recognize why neural networks are important for modern language models.

6.1 The Inspiration from the Human Brain

The human brain is one of the most complex systems known in nature. It contains billions of specialized cells called **neurons** that communicate with one another through electrical and chemical signals. Each neuron receives information from other neurons, processes that information, and sends signals to additional neurons.

This network of connections allows the brain to perform extraordinary tasks such as recognizing faces, understanding language, solving problems, and learning from experience.

Scientists studying artificial intelligence became interested in how the brain processes information. They began asking an important question:

Could computers be designed to learn in a way that resembles how the brain learns?

This question led to the development of **artificial neural networks**, a type of computational system inspired by the structure and behavior of biological neurons.

Although artificial neural networks are much simpler than the human brain, they use similar ideas about interconnected units working together to process information.

6.2 What Is a Neural Network?

A neural network is a machine learning system made up of many connected processing units called **nodes** or **artificial neurons**. These nodes work together to analyze data and identify patterns.

Each artificial neuron receives information, performs a mathematical calculation, and passes the result to other neurons in the network. By combining the outputs of many neurons, the network can learn complex relationships within data.

Neural networks are organized into several layers that help process information step by step.

The first layer receives input data. This input might include numbers representing words, images, or other forms of information. The input is then passed through one or more intermediate layers, often called **hidden layers**, where the network analyzes relationships and patterns. Finally, the information reaches an output layer that produces the system's prediction or result.

Through this layered structure, neural networks are able to transform raw data into meaningful outputs.

6.3 Learning Through Training

Neural networks learn through a process known as **training**. During training, the network analyzes large amounts of data and gradually adjusts its internal parameters in order to improve its predictions.

For example, suppose a neural network is learning to predict the next word in a sentence. The system receives many examples of sentences from a dataset. For each sentence, the network attempts to predict the next word.

At first, the network's predictions may be incorrect. When this happens, the system calculates the difference between its prediction and the correct answer. This difference is called an **error**.

The network then adjusts its internal parameters to reduce the error. Over time, repeated training helps the system improve its ability to make accurate predictions.

This process allows neural networks to gradually learn patterns within the training data.

6.4 Layers and Connections

One of the key features of neural networks is the presence of multiple layers of connected neurons.

Each connection between neurons has an associated numerical value called a **weight**. These weights determine how strongly one neuron influences another.

During training, the network adjusts these weights to improve its predictions. Connections that help produce accurate results become stronger, while less useful connections become weaker.

As information moves through the layers of the network, it is transformed through many mathematical operations. These transformations allow the network to recognize complex patterns that may not be visible in the original data.

For example, when analyzing language, neural networks may learn relationships between words, grammar structures, and contextual meanings.

The deeper the network—that is, the more layers it contains—the more complex patterns it can potentially learn.

6.5 Neural Networks and Language Processing

Neural networks have become one of the most powerful tools for processing human language.

When language data is converted into numerical form, neural networks can analyze relationships between words and phrases across large datasets. Through training, the system learns how words typically appear together and how sentences are structured.

This learning process allows neural networks to perform many language-related tasks, including translation, summarization, question answering, and text generation.

Modern large language models rely heavily on neural networks to interpret language input and produce responses.

Many widely used AI systems that assist users with writing, research, and conversation are built using neural network technology, including systems such as ChatGPT, Claude, and Gemini.

These systems use neural networks with billions of adjustable parameters that allow them to capture extremely detailed patterns in language.

6.6 The Importance of Neural Networks in Modern AI

Neural networks represent a major advancement in artificial intelligence because they allow machines to learn from data rather than relying entirely on human-designed rules.

This capability has enabled significant improvements in many fields, including computer vision, speech recognition, and natural language processing.

In language technologies, neural networks have made it possible for machines to generate text that appears natural and coherent. They allow AI systems to analyze complex relationships between words, topics, and contexts.

Despite these impressive capabilities, it is important to remember that neural networks do not truly understand language the way humans do. Instead, they rely on mathematical patterns learned from large datasets.

Understanding how neural networks function helps students recognize both the strengths and the limitations of modern artificial intelligence.

Chapter Summary

Neural networks are machine learning systems inspired by the structure and function of the human brain. They consist of interconnected processing units that work together to analyze data and identify patterns.

These systems are organized into layers that transform input data into predictions or outputs. During training, neural networks adjust internal parameters in order to improve their accuracy.

In language processing tasks, neural networks analyze numerical representations of words and sentences to learn relationships between language elements.

This ability to learn patterns from large datasets has made neural networks a central technology in modern artificial intelligence.

Large language models rely heavily on neural network architectures to interpret text and generate responses. While these systems can produce impressive results, they operate through statistical learning rather than true human understanding.

In the next chapter, we will explore the development of **transformer architectures**, a breakthrough technology that made modern large language models possible.

Chapter 6 Summary

Neural Networks and Language Learning: How Machines Learn from Data

Neural networks are a key technology behind modern artificial intelligence and large language models. Inspired by the structure of the human brain, neural networks consist of interconnected processing units that work together to analyze information and detect patterns in data.

These systems are organized into layers. The input layer receives information, hidden layers analyze relationships within the data, and the output layer produces a prediction or response. As information moves through the network, mathematical calculations transform the input into meaningful results.

Neural networks learn through a process called training. During training, the system analyzes large datasets and attempts to make predictions. When the prediction is incorrect, the network measures the error and adjusts its internal parameters to improve accuracy. Over time, this process allows the system to recognize patterns and make better predictions.

In language processing tasks, neural networks study numerical representations of words and sentences. By analyzing these patterns across large amounts of text, the system learns how words relate to one another and how sentences are structured.

Modern large language models rely heavily on neural networks to generate responses, translate languages, summarize information, and answer questions. While these systems can produce impressive results, they rely on mathematical pattern recognition rather than true human understanding.

Understanding neural networks helps students appreciate how machines learn from data and why this technology plays such an important role in modern artificial intelligence.

Reflection Section

Chapter 6 – Neural Networks and Language Learning

1. Why were scientists inspired by the human brain when designing neural networks?
2. How do neural networks learn from mistakes during training?
3. Why are layers important in a neural network system?
4. How do neural networks help machines analyze language patterns?
5. Why is it important to understand that neural networks recognize patterns rather than truly understanding language?

Reflection Activity

Write a short paragraph explaining how learning from mistakes helps humans improve their skills. Then compare this idea to how neural networks adjust their parameters during training to improve predictions.

Worksheet

Chapter 6 – Neural Networks and Language Learning

Part 1: Vocabulary Match

Term	Definition
Neural Network	A. A system of interconnected computational units that learn patterns from data
Node	B. A small processing unit within a neural network
Training	C. The process of teaching a machine learning system using data
Layer	D. A stage in a neural network where data is processed

Part 2: Fill in the Blank

1. Artificial neural networks are inspired by the structure of the human __________.
2. The process of teaching a neural network using data is called __________.
3. A small processing unit in a neural network is called a __________.
4. Neural networks learn by adjusting internal __________ during training.
5. Language models analyze patterns in __________ to generate responses.

Part 3: Short Answer

1. Explain how neural networks are similar to the human brain.
2. What happens when a neural network makes an incorrect prediction?

3. Why are large datasets important when training neural networks?
4. How do neural networks help language models understand relationships between words?

Part 4: Critical Thinking

Imagine you are designing a neural network to help students learn new languages.

Write 4–5 sentences explaining how the system might use training data and pattern recognition to help students practice vocabulary and grammar.

Quiz

Chapter 6 – Neural Networks and Language Learning

Multiple Choice

1. What inspired the design of artificial neural networks?
   A. Mathematical formulas only
   B. The structure of the human brain
   C. Computer keyboards
   D. Internet websites

2. What is the main purpose of training a neural network?
   A. To store images
   B. To help the system learn patterns in data
   C. To create computer hardware
   D. To design software programs

3. What is a node in a neural network?
   A. A storage device
   B. A processing unit that receives and sends information
   C. A computer program
   D. A data file

4. What happens when a neural network makes an incorrect prediction during training?
   A. The system stops learning

B. The system deletes the dataset
C. The network adjusts internal parameters to reduce error
D. The system shuts down

5. Why are neural networks important for large language models?
   A. They help machines learn patterns in language data
   B. They replace computer processors
   C. They eliminate the need for datasets
   D. They store internet websites

Short Answer

6. Why do neural networks require large datasets during training?
7. What is one advantage of using neural networks for language processing?

Answer Key

Multiple Choice

1. **B** — The structure of the human brain
2. **B** — To help the system learn patterns in data
3. **B** — A processing unit that receives and sends information
4. **C** — The network adjusts internal parameters to reduce error
5. **A** — They help machines learn patterns in language data

Fill in the Blank

1. **brain**
2. **training**
3. **node**
4. **parameters**
5. **language**

Short Answer (Example Responses)

6. Large datasets provide many examples that allow neural networks to learn patterns more accurately.
7. Neural networks can identify complex relationships between words and sentences, which helps AI systems generate meaningful language responses.

Chapter 7 — Transformer Architecture: The Breakthrough Behind Modern Language Models

Learning Objectives

By the end of this chapter, students will be able to:

1. Explain what transformer architecture is and why it is important for modern language models.
2. Understand how transformer systems process language differently from earlier neural networks.
3. Describe how attention mechanisms help machines understand relationships between words in a sentence.
4. Recognize why transformer models improved the performance of language-processing systems.
5. Explain why transformer architecture made large language models possible.

7.1 The Need for a New Approach

As artificial intelligence systems improved, researchers began developing increasingly complex neural networks to process language. Earlier language models often relied on sequential methods that analyzed words one at a time in a fixed order.

These earlier approaches could process text, but they struggled to understand long sentences or relationships between words that were far apart in a sentence. For example, in a complex paragraph, a key word appearing at the beginning might influence the meaning of a word several sentences later. Earlier models had difficulty capturing these long-distance relationships.

Researchers recognized that a new approach was needed—one that could examine all parts of a sentence simultaneously and understand how words influence one another across an entire passage.

This challenge led to the development of **transformer architecture**, a new type of neural network design that dramatically improved language understanding in artificial intelligence systems.

7.2 What Is Transformer Architecture?

Transformer architecture is a neural network design created specifically to process sequences of information, such as sentences and paragraphs. Instead of analyzing words one by one in strict order, transformer models examine relationships between words across an entire sentence at the same time.

This ability allows the system to recognize how different words interact with each other in a broader context.

The core idea behind transformers is that understanding language requires more than recognizing individual words. Machines must also understand how words relate to each other within a sentence or paragraph.

For example, consider the sentence:

"The scientist who won the award published her research last year."

In this sentence, the word **"scientist"** is closely connected to the pronoun **"her,"** even though several words appear between them. A transformer-based system can recognize this relationship more easily than earlier models.

By analyzing all parts of the sentence simultaneously, transformer systems develop a more complete understanding of language structure.

7.3 The Role of Attention

One of the most important innovations in transformer architecture is a concept known as **attention**.

Attention mechanisms allow a language model to determine which words in a sentence are most relevant when interpreting meaning. Rather than treating every word equally, the model assigns different levels of importance to different words.

For example, in the sentence:

"The teacher explained the lesson because the students were confused."

The word **"because"** signals a relationship between the explanation and the students' confusion. An attention mechanism helps the model focus on this relationship when interpreting the sentence.

In practice, attention mechanisms allow the model to examine how each word relates to every other word in the sentence. This process helps the system identify grammatical relationships, topic connections, and contextual meaning.

Because attention enables the model to focus on relevant information, it significantly improves the system's ability to understand complex language.

7.4 Parallel Processing of Language

Another important advantage of transformer architecture is its ability to process information in parallel.

Earlier neural network models often analyzed sentences step by step, processing one word after another. This approach required significant time and computing resources, especially when working with long passages of text.

Transformers take a different approach. They analyze multiple parts of a sentence simultaneously, allowing the model to process large amounts of language data more efficiently.

This parallel processing capability made it possible to train language models on extremely large datasets containing billions of words.

As a result, transformer models can capture patterns and relationships in language that earlier systems could not easily detect.

7.5 Scaling Language Models

Transformer architecture made it possible to build much larger and more powerful language models.

Because transformer systems can process large amounts of data efficiently, researchers began training models with increasing numbers of parameters and larger datasets. These expanded models became capable of generating more accurate and coherent text.

The ability to scale language models led to the creation of modern large language models that can perform many different tasks, including answering questions, summarizing documents, translating languages, and assisting with writing.

Many widely used AI systems today rely on transformer-based architectures, including systems such as ChatGPT, Claude, and Gemini.

These systems demonstrate how transformer models can generate detailed responses and assist users in a variety of educational and professional tasks.

7.6 Why Transformers Changed Artificial Intelligence

The introduction of transformer architecture marked a major turning point in artificial intelligence research. By allowing machines to analyze language relationships across entire passages, transformers greatly improved the accuracy and flexibility of language models.

This breakthrough enabled researchers to build systems capable of generating natural-sounding text and engaging in extended conversations with users.

Transformers also made it possible to train models on much larger datasets than before. With access to more data and more powerful computing resources, language models could learn increasingly sophisticated patterns in language.

As a result, transformer-based systems became the foundation for many of the most advanced AI technologies used today.

Understanding transformer architecture helps students appreciate how modern language models operate and why this technology represents such an important advancement in artificial intelligence.

Chapter Summary

Transformer architecture represents one of the most significant breakthroughs in modern artificial intelligence. Unlike earlier neural network models that processed language sequentially, transformers analyze relationships between words across an entire sentence or passage at the same time.

The introduction of attention mechanisms allows transformer models to focus on the most relevant words when interpreting meaning. This ability improves the system's understanding of context and complex language structures.

Transformers also support parallel processing, enabling models to analyze large amounts of text efficiently. This capability made it possible to train language models on massive datasets and scale them to unprecedented sizes.

Because of these advantages, transformer architecture became the foundation for modern large language models. These systems are now used across many industries to assist with communication, research, education, and creative work.

In the next chapter, we will explore **how large language models are trained**, examining the data, computing power, and learning processes that allow these systems to develop their language capabilities.

Chapter Summary

Chapter 7 – Transformer Architecture: The Breakthrough Behind Modern Language Models

Transformer architecture represents one of the most important technological breakthroughs in the development of modern artificial intelligence systems. Earlier language models often processed sentences sequentially, analyzing words one at a time. While these models could recognize patterns in text, they struggled to capture long-distance relationships between words and required significant computing time when processing large amounts of data.

Transformer models introduced a new approach to language processing by analyzing relationships between words across an entire sentence simultaneously. This architecture allows the model to consider the full context of a sentence rather than focusing on words individually in sequence.

A key component of transformer architecture is the attention mechanism. Attention allows the model to identify which words are most important when interpreting meaning. By focusing on relevant parts of a sentence, the system can better understand grammatical structure and contextual relationships.

Another advantage of transformers is their ability to process information in parallel. Instead of analyzing words one by one, transformer systems examine multiple parts of the text at the same time. This capability allows the model to process large datasets more efficiently and train on massive collections of language data.

Because transformers can process large datasets and analyze complex relationships between words, they became the foundation for modern large language models. These systems are now used in many applications such as writing assistance, language translation, research support, and educational tools.

Understanding transformer architecture helps students recognize how advances in computing and machine learning have enabled modern artificial intelligence systems to generate and interpret language with increasing sophistication.

Reflection Section

Chapter 7 – Transformer Architecture

1. Why were earlier language-processing systems limited in their ability to understand long sentences?
2. How does transformer architecture improve language processing compared to earlier neural network models?
3. Why is the attention mechanism important for understanding relationships between words in a sentence?
4. How does parallel processing make transformer models more efficient?
5. Why is transformer architecture considered a major breakthrough in artificial intelligence?

Reflection Activity

Write a short paragraph explaining how focusing attention on important details can improve understanding. You may use an example from reading, studying, or problem-solving and compare it to how attention mechanisms help AI models interpret language.

Worksheet

Chapter 7 – Transformer Architecture

Part 1: Vocabulary Match

Term	Definition
Transformer Architecture	A. A neural network design that analyzes relationships between words in language
Attention Mechanism	B. A system that helps AI focus on the most relevant words in a sentence
Parallel Processing	C. The ability to analyze multiple pieces of information at the same time
Context	D. The surrounding words that influence the meaning of a sentence

Part 2: Fill in the Blank

1. Transformer architecture allows AI systems to analyze __________ between words across a sentence.
2. The __________ mechanism helps the model focus on important words when interpreting meaning.
3. __________ processing allows models to analyze multiple parts of a sentence at the same time.
4. Earlier language models processed sentences __________, meaning one word at a time.
5. Transformer models improved the performance of modern __________ models.

Part 3: Short Answer

1. Explain why earlier language models struggled with long sentences.
2. What role does the attention mechanism play in transformer models?
3. How does parallel processing improve efficiency in language models?
4. Why did transformer architecture make it possible to build larger language models?

Part 4: Critical Thinking

Imagine you are designing a system that reads long scientific articles and summarizes them.

Write 4–5 sentences explaining how transformer architecture and attention mechanisms could help the system identify important information in the article.

Quiz

Chapter 7 – Transformer Architecture

Multiple Choice

1. What is the main advantage of transformer architecture?
   A. It stores data more efficiently
   B. It analyzes relationships between words across a sentence
   C. It replaces neural networks
   D. It reduces the need for datasets

2. What does an attention mechanism help a language model do?
   A. Store text files
   B. Focus on important words in a sentence
   C. Translate images
   D. Increase computer speed

3. What does parallel processing allow transformer models to do?
   A. Analyze information sequentially
   B. Process multiple parts of text at the same time
   C. Store larger datasets
   D. Eliminate neural networks

4. Why did transformers allow the development of large language models?
   A. They reduced the amount of data required
   B. They could process large datasets efficiently
   C. They removed the need for training
   D. They eliminated algorithms

5. Transformer models are primarily used to process what type of information?
   A. Mechanical systems
   B. Numerical calculations only
   C. Human language and sequences of text
   D. Electrical circuits

Short Answer

6. What is one reason transformer architecture improved language understanding?
7. Why is context important when interpreting language?

Answer Key

Multiple Choice

1. **B** — It analyzes relationships between words across a sentence
2. **B** — Focus on important words in a sentence
3. **B** — Process multiple parts of text at the same time
4. **B** — They could process large datasets efficiently
5. **C** — Human language and sequences of text

Fill in the Blank

1. **relationships**
2. **attention**
3. **Parallel**
4. **sequentially**
5. **language**

Short Answer (Example Responses)

6. Transformer architecture improved language understanding by allowing models to analyze relationships between words across an entire sentence rather than processing words one at a time.
7. Context helps determine the correct meaning of words because many words can have different meanings depending on the surrounding sentence.

Chapter 8 — Training Large Language Models: How Machines Learn Language

Learning Objectives

By the end of this chapter, students will be able to:

1. Explain what it means to train a large language model.
2. Describe the role of datasets in teaching machines how language works.
3. Understand how models learn patterns through repeated training cycles.
4. Recognize the importance of computing power in training modern AI systems.

5. Explain why training data and learning processes influence the behavior of language models.

8.1 What Does It Mean to Train an AI Model?

Training is the process through which a machine learning system learns patterns from data. For large language models, training involves analyzing vast collections of text so that the system can learn how words and sentences are structured.

Unlike traditional computer programs that follow fixed instructions, language models improve their performance by studying examples. During training, the model repeatedly examines text and attempts to predict the next word or token in a sequence.

When the model makes an incorrect prediction, the system adjusts its internal parameters to reduce the error. Over time, these adjustments allow the model to recognize patterns in language more accurately.

This process is similar to how humans learn through practice and feedback. Just as students improve their understanding by studying examples and correcting mistakes, language models refine their predictions through repeated exposure to data.

8.2 The Role of Training Data

The knowledge of a language model comes primarily from the data it studies during training. This training data usually consists of enormous collections of written language gathered from many sources.

Examples of training data may include books, articles, scientific papers, and other forms of digital text. By analyzing large datasets, the model observes how words appear together, how sentences are structured, and how ideas develop within paragraphs.

The variety and quality of training data play an important role in determining how well a model performs. If a dataset contains diverse and well-structured information, the model may develop a stronger understanding of language patterns.

However, if the training data contains errors or biases, the model may learn patterns that reflect those problems. For this reason, researchers must carefully evaluate the data used during training.

Training data therefore shapes the abilities and limitations of language models.

8.3 Predicting the Next Word

One of the fundamental tasks used to train language models is predicting the next word in a sentence.

For example, consider the partial sentence:

"The sun rises in the..."

A language model trained on large datasets may predict that the next word is likely to be *"east."*

To make this prediction, the system calculates probabilities for many possible words and selects the one most likely to follow the previous words in the sequence.

During training, the model performs millions or even billions of these prediction tasks. Each prediction helps the system refine its understanding of language patterns.

Over time, these repeated learning steps allow the model to generate longer passages of text that appear coherent and meaningful.

8.4 The Importance of Computing Power

Training large language models requires enormous computing resources. Because these systems analyze extremely large datasets and contain billions of adjustable parameters, the training process involves vast numbers of mathematical calculations.

To perform these calculations efficiently, researchers use specialized hardware designed for high-performance computing. These systems allow models to process massive amounts of data during training.

The availability of powerful computing infrastructure has been one of the key factors enabling the development of modern AI systems.

Without sufficient computing power, training a large language model could take many years. Advanced hardware systems allow researchers to complete training in a much shorter period of time.

8.5 Learning Through Iteration

Training a language model is not a single event but a repeated cycle of improvement.

During each training step, the model performs a prediction, measures the error, and adjusts its parameters. These adjustments gradually improve the model's ability to generate accurate predictions.

This process continues for many training cycles, sometimes called **iterations**. Each iteration allows the model to refine its understanding of language patterns.

As training progresses, the model becomes increasingly capable of generating coherent sentences, interpreting questions, and summarizing information.

Because training involves many iterations over large datasets, it allows the model to capture complex relationships between words and ideas.

8.6 From Training to Real-World Use

After training is completed, the language model can be used to perform a variety of tasks. These tasks may include answering questions, assisting with writing, translating languages, and summarizing documents.

Many modern AI tools rely on trained language models to interact with users in natural language. Systems such as ChatGPT, Claude, and Gemini demonstrate how trained models can assist users in communication and learning.

However, it is important to remember that the responses generated by these systems depend heavily on the patterns learned during training. Language models do not possess human awareness or understanding; they generate responses based on statistical patterns found in their training data.

Understanding how models are trained helps students appreciate both the capabilities and the limitations of modern AI systems.

Chapter Summary

Training is the process through which large language models learn patterns in human language. By analyzing massive datasets, these systems learn how words and sentences are structured.

During training, the model repeatedly attempts to predict the next word in a sequence. When predictions are incorrect, the system adjusts its internal parameters to reduce error. Over many iterations, this process allows the model to refine its understanding of language patterns.

The availability of large datasets and powerful computing resources has made it possible to train increasingly sophisticated language models. These systems can perform a wide range of language-related tasks, including answering questions, summarizing information, and assisting with writing.

Although modern language models can generate impressive responses, they rely entirely on statistical patterns learned during training. They do not possess human understanding or awareness.

In the next chapter, we will explore the concept of **parameters**, examining how these adjustable values store the knowledge that language models learn during training.

Reflection Section

Chapter 8 – Training Large Language Models: How Machines Learn Language

Reflection questions encourage students to think about how artificial intelligence systems learn and how this learning process compares with human learning.

1. Why is training necessary for large language models to function effectively?
2. How is the process of training an AI model similar to the way humans learn through practice?
3. Why does the quality of training data influence how well a language model performs?
4. Why do large language models require powerful computing systems during training?
5. Why is it important to understand that AI models learn patterns rather than truly understanding language?

Reflection Activity

Write a short paragraph explaining how repeated practice helps people learn new skills such as mathematics, sports, or music. Then compare this process to how large language models improve their predictions through repeated training cycles.

Worksheet

Chapter 8 – Training Large Language Models

Part 1: Vocabulary Match

Match each term with the correct definition.

Term	Definition
Training	A. The process of teaching a machine learning model using data
Dataset	B. A large collection of information used to train AI systems
Iteration	C. A repeated cycle used to improve a model's predictions
Prediction	D. An estimate of what word or outcome is most likely to occur

Part 2: Fill in the Blank

1. The process of teaching a machine learning system using data is called __________.
2. A large collection of information used to train AI systems is called a __________.
3. Language models learn by predicting the next __________ in a sequence.
4. Each repeated training cycle is called an __________.
5. Powerful __________ systems are required to train large language models efficiently.

Part 3: Short Answer

1. Explain why large datasets are important when training language models.

2. How does predicting the next word help a model learn language patterns?
3. Why do language models require repeated training cycles?
4. What role does computing power play in training modern AI systems?

Part 4: Critical Thinking

Imagine you are responsible for training a language model designed to help students with homework.

Write 4–5 sentences explaining what types of training data you would include and why this data would help the model assist students effectively.

Quiz

Chapter 8 – Training Large Language Models

Multiple Choice

1. What does training mean in machine learning?
   A. Writing computer programs manually
   B. Teaching a model to learn patterns from data
   C. Installing computer software
   D. Storing information in a database

2. What is a dataset used for in AI training?
   A. To build computer hardware
   B. To provide examples that the model learns from
   C. To store internet websites
   D. To control computer speed

3. What is one of the main tasks used to train language models?
   A. Generating images
   B. Predicting the next word in a sentence
   C. Translating videos
   D. Designing computer chips

4. Why do large language models require powerful computers during training?
   A. To store images
   B. To perform many mathematical calculations quickly
   C. To connect to the internet
   D. To create new programming languages

5. What happens when a language model makes an incorrect prediction during training?
   A. The system stops working
   B. The model adjusts its parameters to reduce error
   C. The dataset is deleted
   D. The program shuts down

Short Answer

6. Why is repetition important in the training process of a language model?
7. How can the quality of training data affect the performance of an AI system?

Answer Key

Multiple Choice

1. **B** — Teaching a model to learn patterns from data
2. **B** — To provide examples that the model learns from
3. **B** — Predicting the next word in a sentence
4. **B** — To perform many mathematical calculations quickly
5. **B** — The model adjusts its parameters to reduce error

Fill in the Blank

1. **training**
2. **dataset**
3. **word** (or **token**)
4. **iteration**
5. **computing**

Short Answer (Example Responses)

6. Repetition allows the model to gradually improve its predictions by adjusting internal parameters over many training cycles.
7. High-quality and diverse training data helps the model learn accurate language patterns, while poor or biased data may lead to incorrect or unfair predictions.

Chapter 9 — Parameters: How Language Models Store Knowledge

Learning Objectives

By the end of this chapter, students will be able to:

1. Explain what parameters are in a large language model.
2. Describe how parameters store patterns learned during training.
3. Understand how adjusting parameters allows AI systems to improve predictions.
4. Recognize why large language models contain billions or trillions of parameters.
5. Explain how parameters influence the behavior and capabilities of language models.

9.1 What Are Parameters?

In machine learning, **parameters** are the adjustable numerical values inside a model that determine how the system processes information.

These values function as the internal memory of the model. During training, the system modifies these numbers in order to improve its predictions.

Every time a model analyzes text and attempts to predict the next word, it uses its parameters to calculate probabilities for possible outcomes. If the prediction is incorrect, the model adjusts the parameters slightly so that future predictions become more accurate.

Over many training cycles, these adjustments allow the system to capture patterns in language.

Although parameters are simply numbers within a mathematical model, they collectively represent the knowledge the system has learned from its training data.

9.2 Learning Through Parameter Adjustment

When a language model is first created, its parameters are usually set to random values. At this stage, the model has no understanding of language patterns.

During training, the system processes large amounts of text and repeatedly attempts to predict the next word in a sequence. After each prediction, the model compares its output with the correct answer.

If the prediction is incorrect, the model calculates the error and adjusts its parameters to reduce that error. This process allows the system to gradually refine the numerical values inside the network.

Through repeated adjustments, the model learns relationships between words, sentence structures, and contextual patterns.

The more training the model receives, the more refined its parameters become.

9.3 Parameters as Encoded Knowledge

Parameters can be thought of as the way a model stores the knowledge it learns from data.

However, this knowledge is not stored in the form of sentences or explicit facts. Instead, it is stored as numerical relationships between words and patterns.

For example, after training on many sentences, a model may learn that the words *doctor*, *hospital*, and *patient* often appear in related contexts. The parameters of the model encode these relationships through mathematical weights and connections.

Because of this structure, the model does not retrieve information in the same way a database does. Instead, it generates responses based on patterns embedded in its parameters.

This difference is important when understanding how large language models produce answers.

9.4 The Scale of Modern Language Models

One of the defining characteristics of modern large language models is the enormous number of parameters they contain.

Early neural networks contained relatively small numbers of adjustable parameters. As computing power increased and training data expanded, researchers began developing models with far larger parameter counts.

Modern language models may contain billions or even trillions of parameters. Each parameter represents a small numerical adjustment that helps the model capture subtle patterns in language.

A larger number of parameters allows the model to represent more complex relationships between words and ideas.

However, larger models also require more computing resources for training and operation. Managing these large systems requires advanced computing infrastructure and careful model design.

9.5 Parameters and Model Behavior

The parameters of a language model determine how the system responds to input.

Because parameters encode patterns learned from training data, they influence how the model interprets language, generates responses, and connects ideas.

If a model has been trained on diverse and well-structured data, its parameters may allow it to generate more accurate and useful responses. If the training data contains errors or biases, those patterns may also influence the model's behavior.

For this reason, researchers carefully evaluate both the data used for training and the models produced from that data.

Understanding parameters helps explain why language models sometimes generate surprising or unexpected responses.

9.6 Parameters in Modern AI Systems

Large language models used in modern AI systems rely on extremely large parameter networks to process language.

These parameters enable systems to analyze relationships between words, detect contextual meaning, and generate complex responses to user questions.

Many widely used AI tools are powered by language models with billions of parameters, including systems such as ChatGPT, Claude, and Gemini.

These systems demonstrate how large parameter networks allow machines to generate text, answer questions, and assist with learning tasks.

Despite their impressive capabilities, these models still rely entirely on patterns encoded within their parameters rather than human-like understanding.

Recognizing this distinction helps students develop a clearer perspective on how artificial intelligence operates.

Chapter Summary

Parameters are the adjustable numerical values inside a machine learning model that determine how the system processes information.

During training, the model repeatedly analyzes text and adjusts its parameters to reduce prediction errors. Through this process, the model gradually learns patterns in language.

Although parameters are simply numbers, they encode relationships between words and concepts that allow the model to generate meaningful responses.

Modern large language models contain billions or even trillions of parameters, enabling them to capture extremely complex patterns in language data.

Understanding how parameters store learned knowledge helps explain how language models generate responses and why their behavior depends heavily on the data used during training.

In the next chapter, we will explore how language models interpret **context**, examining how AI systems determine meaning by analyzing relationships between words within sentences and conversations.

Chapter 9 Summary

Parameters: How Language Models Store Knowledge

Parameters are the adjustable numerical values inside a machine learning model that determine how the system processes information. They function as the internal structure that allows a language model to learn patterns from data. During training, the model repeatedly analyzes text and attempts to predict the next word in a sentence. When the prediction is incorrect, the system adjusts its parameters slightly to reduce the error.

Over time, these adjustments allow the model to capture relationships between words, phrases, and ideas. Instead of storing information like a database, language models encode patterns within their parameters through mathematical relationships.

Modern large language models contain billions or even trillions of parameters. These massive networks allow AI systems to recognize complex language structures and generate detailed responses. However, the knowledge stored in parameters depends entirely on the data used during training. Understanding parameters helps students recognize how AI models learn and why their responses reflect patterns learned from data rather than true human understanding.

Reflection Section

Chapter 9 – Parameters

Reflection helps students connect technical concepts with broader ideas about learning and intelligence.

1. Why are parameters important in machine learning systems?
2. How do parameters change during the training process?
3. Why do modern language models contain billions of parameters?
4. How is the knowledge stored in parameters different from information stored in a traditional database?
5. Why is it important for people to understand that AI models rely on numerical patterns rather than human understanding?

Reflection Activity

Think about how people improve their skills through practice. Write a short paragraph explaining how practice changes the way the brain performs tasks such as reading, playing music, or solving math problems. Then compare this idea with how parameter adjustments help AI systems improve their predictions.

Worksheet

Chapter 9 – Parameters

Part 1: Vocabulary Match

Match the term with its correct definition.

Term	Definition
Parameter	A. Numerical values inside a model that control how it processes information
Training	B. The process of teaching a model using data
Prediction	C. An estimate of the most likely next word or outcome
Model	D. A mathematical system designed to learn patterns from data

Part 2: Fill in the Blank

1. __________ are adjustable numerical values inside a machine learning model.

2. During training, the model attempts to __________ the next word in a sentence.
3. If the model makes a mistake, it adjusts its __________ to reduce the error.
4. Modern language models may contain billions or even __________ of parameters.
5. Parameters allow the system to store __________ learned from data.

Part 3: Short Answer

1. Explain what a parameter is in a language model.
2. Why do parameters change during training?
3. How do parameters help language models recognize patterns in language?
4. Why do large language models require many parameters?

Part 4: Critical Thinking

Imagine you are designing a language model that helps students learn science concepts.

Write 4–5 sentences explaining how the model's parameters would help it understand relationships between words such as *energy*, *matter*, *force*, and *motion*.

Quiz

Chapter 9 – Parameters

Multiple Choice

1. What is a parameter in a machine learning model?
   A. A type of computer hardware
   B. A numerical value that influences how the model processes information
   C. A programming language
   D. A dataset used for training

2. What happens when a model makes an incorrect prediction during training?
   A. The system shuts down
   B. The parameters are adjusted to reduce the error
   C. The dataset is deleted
   D. The model stops learning

3. Why do modern language models contain billions of parameters?
   A. To store pictures
   B. To represent complex language patterns
   C. To reduce computer memory
   D. To simplify programming

4. How do language models store knowledge?
   A. As written sentences in memory
   B. As mathematical relationships encoded in parameters
   C. As saved internet websites
   D. As lists of vocabulary words

5. What determines how a language model behaves?
   A. Its color and design
   B. The number of users
   C. The parameters learned during training
   D. The operating system

Short Answer

6. Why are parameters essential for machine learning systems?
7. How can the quality of training data influence the parameters learned by a model?

Answer Key

Multiple Choice

1. **B** — A numerical value that influences how the model processes information
2. **B** — The parameters are adjusted to reduce the error
3. **B** — To represent complex language patterns
4. **B** — As mathematical relationships encoded in parameters
5. **C** — The parameters learned during training

Fill in the Blank

1. **Parameters**
2. **predict**
3. **parameters**
4. **trillions**
5. **patterns**

Short Answer (Example Responses)

6. Parameters are essential because they determine how the model processes information and allows it to learn patterns from data.
7. High-quality training data helps the model learn accurate language relationships, while poor or biased data can lead to incorrect or misleading patterns.

Chapter 10 — Context: How Language Models Understand Meaning

Learning Objectives

By the end of this chapter, students will be able to:

1. Explain what context means in human language and communication.
2. Describe how large language models interpret words based on surrounding text.
3. Understand why context is necessary for determining the correct meaning of words.
4. Recognize how context allows language models to generate coherent responses.
5. Identify limitations that occur when AI systems misunderstand context.

10.1 What Is Context?

Language is more than individual words. The meaning of a sentence often depends on the relationship between words, ideas, and situations. This surrounding information is known as **context**.

Context includes the words that appear before and after a particular term, the subject being discussed, and the overall structure of a conversation. Humans naturally use context to interpret meaning. When people hear or read a sentence, their brains quickly analyze surrounding clues to understand what the speaker intends.

For example, the word *bank* can have different meanings depending on context. In one sentence it may refer to a financial institution, while in another it may describe the edge of a river.

Humans determine the correct meaning by examining the surrounding words and the topic of discussion.

Large language models must also rely on context to interpret language correctly. Without context, words alone would be ambiguous and difficult to interpret.

10.2 Context in Human Communication

Human communication relies heavily on context. When people speak, they often assume that listeners already understand certain background information.

For instance, a teacher may say, "Open your books to page ten." Students know the teacher is referring to a specific textbook used in class. The meaning of the instruction depends on shared knowledge and the environment in which the conversation takes place.

Humans also use emotional and social context when interpreting language. Tone of voice, facial expressions, and cultural expectations influence how messages are understood.

Because humans interpret language through many layers of context, communication can remain flexible and adaptable even when words are brief or incomplete.

10.3 Context in Large Language Models

Large language models analyze context by examining the sequence of tokens in a sentence or conversation. Instead of interpreting emotional cues or real-world experiences, the model calculates relationships between words based on patterns learned during training.

When a user writes a prompt, the model studies the words that appear before and after each token in the sentence. Using its learned parameters, it estimates which meanings and responses are most likely within that context.

For example, if a sentence includes the words *river*, *water*, and *fishing*, the model is more likely to interpret the word *bank* as the side of a river rather than a financial institution.

This ability to analyze surrounding information allows the model to generate responses that appear logical and coherent.

10.4 Context Windows

Large language models operate within what researchers call a **context window**. A context window is the amount of text the model can consider at one time when generating a response.

If the context window is large, the model can analyze longer passages of text and maintain coherence across extended conversations or documents. Smaller context windows limit how much information the system can use when generating responses.

Advances in AI research have increased the size of context windows in modern language models. Larger context windows allow systems to summarize long articles, maintain detailed conversations, and analyze complex documents.

However, even the largest context windows have limits. When conversations exceed those limits, earlier parts of the discussion may no longer influence the model's responses.

10.5 Understanding Meaning Through Patterns

Although language models appear to understand language, they actually rely on statistical patterns within text.

The model does not experience the world the way humans do. It does not see, hear, or feel events directly. Instead, it learns relationships between words through patterns found in training data.

When analyzing context, the model calculates probabilities for possible responses based on these learned patterns. This statistical process allows the system to generate meaningful text, even though it does not possess personal experiences or emotions.

Recognizing this distinction helps students understand both the strengths and limitations of artificial intelligence.

10.6 Context and Coherent Conversations

One of the most important uses of context is maintaining coherent conversations. When users interact with AI systems, the model must track the topic of discussion and generate responses that relate to previous statements.

By analyzing earlier messages in the conversation, the model can maintain continuity and respond in ways that appear thoughtful and relevant.

For example, if a conversation begins with a discussion about space exploration, the model will continue generating responses related to astronomy, spacecraft, or scientific discovery.

Without context, the model would generate unrelated responses that would appear confusing or random.

10.7 When Context Fails

Despite their abilities, language models can sometimes misunderstand context.

Because they rely on statistical patterns rather than real-world understanding, they may occasionally misinterpret ambiguous sentences or complex instructions. When context is unclear or incomplete, the model may generate responses that do not match the user's intention.

Researchers continue working to improve how language models interpret context, especially in complex conversations and technical subjects.

Students studying AI should recognize that even advanced systems require careful human guidance to ensure accurate interpretation.

10.8 Context in Modern AI Systems

Many modern AI tools rely on contextual language processing to interact with users. Systems such as ChatGPT, Claude, and Gemini demonstrate how context-based language models can assist users with writing, research, learning, and communication.

These systems analyze the context of user prompts to generate helpful responses. The ability to interpret context is one of the most important factors that allows AI systems to communicate effectively with people.

Understanding how context works helps students develop a clearer perspective on how language models function and why context plays such a critical role in AI-powered communication.

Chapter Summary

Context refers to the surrounding information that helps determine the meaning of words and sentences. In human communication, context includes prior knowledge, emotional cues, and environmental factors that influence how language is interpreted.

Large language models analyze context by examining relationships between words in a sequence of tokens. Using parameters learned during training, the model calculates probabilities to determine which responses are most appropriate within a given context.

The concept of a context window determines how much text a model can consider at one time. Larger context windows allow models to maintain coherence across longer conversations and documents.

Although language models can analyze context effectively, they rely on statistical patterns rather than true human understanding. Recognizing both the capabilities and limitations of context-based AI systems helps students develop a deeper understanding of how modern language models operate.

In the next chapter, we will explore **fine-tuning and alignment**, examining how researchers refine language models after training to improve accuracy, safety, and usefulness.

Chapter 10 Summary

Context: How Language Models Understand Meaning

Context is the surrounding information that helps determine the meaning of words, sentences, and conversations. In human communication, context includes the words around a statement, the situation in which the conversation occurs, and the shared knowledge between speakers.

Large language models also rely on context to interpret language. Instead of using emotions or personal experiences, these models analyze the sequence of words in a sentence and identify patterns learned during training. By examining how words relate to one another, the model can estimate the most likely meaning and generate an appropriate response.

Language models operate within a **context window**, which determines how much text the system can analyze at one time. A larger context window allows the model to consider longer passages of information and maintain coherence during extended conversations.

Although AI systems can interpret context effectively in many situations, they still rely on statistical patterns rather than true understanding. Recognizing how context works helps students understand both the strengths and limitations of large language models.

Reflection Section

Chapter 10 – Context

Reflection encourages students to think deeply about how meaning is created in communication.

1. Why is context important when interpreting the meaning of words?
2. How do humans use context differently from AI systems?
3. Why can a single word have multiple meanings depending on context?
4. What role does a context window play in how language models generate responses?
5. Why is it important for users to provide clear prompts when interacting with AI systems?

Reflection Activity

Write a short paragraph describing a situation in which a word or sentence could have multiple meanings depending on context. Explain how surrounding information helps people determine the correct meaning and compare this to how language models analyze context in text.

Worksheet

Chapter 10 – Context

Part 1: Vocabulary Match

Match each term with the correct definition.

Term	Definition
Context	A. Surrounding information that helps determine meaning
Context Window	B. The amount of text a language model can analyze at one time
Token	C. A unit of text used by AI models to process language
Interpretation	D. The process of determining meaning from information

Part 2: Fill in the Blank

1. __________ refers to the surrounding information that helps determine the meaning of language.
2. Large language models analyze sequences of __________ to understand text.
3. A __________ determines how much text a language model can consider at once.
4. Words can have different meanings depending on their __________.
5. Language models generate responses based on patterns learned from __________ data.

Part 3: Short Answer

1. Explain why context is necessary for understanding language.
2. How do large language models analyze context in sentences?
3. What is a context window and why is it important?
4. Why might AI systems misunderstand context in certain situations?

Part 4: Critical Thinking

Imagine a language model is assisting students in an online classroom.

Write 4–5 sentences explaining how context would help the model understand questions related to different subjects such as mathematics, science, or literature.

Quiz

Chapter 10 – Context

Multiple Choice

1. What does context refer to in language?
   A. The number of letters in a word
   B. The surrounding information that determines meaning
   C. The pronunciation of a word
   D. The spelling of a sentence

2. Why is context important in communication?
   A. It helps determine the correct meaning of words
   B. It changes the spelling of words
   C. It reduces the number of sentences
   D. It eliminates grammar rules

3. What does a context window determine in a language model?
   A. The amount of text the model can analyze at one time
   B. The number of datasets used for training
   C. The size of the computer screen
   D. The speed of the internet

4. How do language models interpret context?
   A. Through emotional understanding
   B. Through statistical patterns learned from data
   C. Through personal experience
   D. Through human memory

5. Why might AI systems sometimes misunderstand context?
   A. Because they rely on statistical patterns rather than real-world understanding
   B. Because they cannot process words

C. Because they cannot store information
D. Because they cannot generate sentences

Short Answer

6. What is one example of a word that can have different meanings depending on context?
7. Why do users need to provide clear prompts when interacting with AI systems?

Answer Key

Multiple Choice

1. **B** — The surrounding information that determines meaning
2. **A** — It helps determine the correct meaning of words
3. **A** — The amount of text the model can analyze at one time
4. **B** — Through statistical patterns learned from data
5. **A** — Because they rely on statistical patterns rather than real-world understanding

Fill in the Blank

1. **Context**
2. **tokens**
3. **context window**
4. **context**
5. **training**

Short Answer (Example Responses)

6. Words such as *bank*, *light*, or *charge* can have different meanings depending on context.
7. Clear prompts provide better context for the AI system, helping the model generate more accurate and relevant responses.

Chapter 11 — Fine-Tuning and Alignment: Teaching AI to Be Helpful, Safe, and Accurate

Learning Objectives

By the end of this chapter, students will be able to:

1. Explain what **fine-tuning** means in artificial intelligence.
2. Describe why AI systems must be aligned with human values and goals.
3. Understand how developers improve language models after initial training.
4. Recognize the importance of safety and responsibility in AI systems.
5. Explain how alignment helps ensure that AI systems assist people in positive and useful ways.

11.1 From Training to Refinement

After a large language model completes its initial training, the system has learned many patterns from large datasets. However, the model may still produce responses that are unclear, inaccurate, or inappropriate.

Because of this, researchers continue refining the model after training. This process is known as **fine-tuning**.

Fine-tuning involves adjusting the model using carefully selected examples and feedback. These adjustments help guide the model toward producing responses that are more accurate, helpful, and aligned with human expectations.

While the initial training teaches the model the structure of language, fine-tuning improves how the model interacts with people.

11.2 What Is Alignment?

Alignment refers to the process of ensuring that an artificial intelligence system behaves in ways that reflect human goals, ethical values, and safety standards.

Because AI systems learn patterns from large datasets, they do not automatically understand social norms or moral considerations. Without guidance, a model might generate responses that are misleading or inappropriate.

Alignment helps ensure that AI systems provide information responsibly. Developers work to train models so that they prioritize helpfulness, fairness, and accuracy when responding to users.

In many ways, alignment functions as a form of guidance that helps shape how AI systems behave in real-world applications.

11.3 Human Feedback in AI Training

One important method used in fine-tuning language models involves **human feedback**.

During this process, human reviewers examine responses generated by the model and evaluate their quality. Reviewers may compare multiple responses and identify which one is more helpful, accurate, or appropriate.

These evaluations help researchers adjust the model so that it produces responses similar to the ones humans considered best.

Through repeated cycles of feedback and adjustment, the system becomes better at generating responses that align with human expectations.

This process highlights an important idea: even highly advanced AI systems still rely on human guidance to improve their behavior.

11.4 Safety and Responsible AI

As artificial intelligence systems become more powerful, ensuring their safe and responsible use becomes increasingly important.

Fine-tuning and alignment help reduce harmful outputs and improve the reliability of AI systems. Researchers work to design models that avoid generating misleading information, harmful instructions, or biased responses.

Developers also create guidelines and safeguards that help ensure AI systems are used in ways that benefit society.

For students learning about artificial intelligence, this topic emphasizes that technology must always be paired with responsibility.

Intelligence alone does not guarantee positive outcomes; ethical guidance is essential.

11.5 Teaching AI to Be Helpful

Alignment is not only about preventing problems; it is also about improving the usefulness of AI systems.

Developers refine models so they can provide clearer explanations, more accurate answers, and better assistance to users.

For example, a well-aligned AI system should be able to explain complex ideas in ways that match a student's level of understanding. It should also recognize when it does not have enough information and respond carefully rather than producing misleading statements.

These improvements help transform language models into practical tools for education, research, and communication.

11.6 Continuous Improvement

The development of AI systems does not end after a model is released. Researchers continue studying how these systems behave and work to improve them over time.

New training methods, updated datasets, and improved alignment techniques help refine AI systems as technology evolves.

This ongoing process reflects an important principle of technology development: intelligent systems must continue learning and adapting to meet the needs of society.

Just as education helps humans grow throughout life, responsible development helps AI systems become more reliable and beneficial tools.

11.7 Alignment in Modern AI Systems

Many modern AI tools rely on alignment techniques to improve their interactions with users. Systems such as ChatGPT, Claude, and Gemini are developed using methods that combine large-scale training with additional refinement and human feedback.

These systems illustrate how alignment helps transform powerful language models into tools designed to assist people responsibly.

Understanding alignment allows students to see that artificial intelligence is not only about technology but also about ethics, responsibility, and thoughtful design.

Chapter 11 Summary

Fine-Tuning and Alignment: Teaching AI to Be Helpful, Safe, and Accurate

Fine-tuning and alignment are important steps in improving artificial intelligence systems after their initial training. During the first stage of training, a large language model learns general language patterns from large datasets. However, the model may still generate responses that are unclear, inaccurate, or inappropriate.

Fine-tuning improves the model by providing additional examples and adjustments that guide the system toward better responses. Alignment focuses on ensuring that AI systems behave in ways that reflect human values, safety standards, and ethical guidelines.

Human feedback plays a key role in this process. Researchers and reviewers evaluate model responses and help train the system to produce answers that are helpful, accurate, and appropriate.

As artificial intelligence becomes more widely used in education, business, and communication, fine-tuning and alignment help ensure that these technologies remain responsible tools that support human learning and decision-making.

Reflection Section

Chapter 11 – Fine-Tuning and Alignment

Reflection helps students consider the relationship between technology and responsibility.

1. Why is fine-tuning necessary after the initial training of an AI model?
2. What does alignment mean in artificial intelligence systems?
3. Why is human feedback important in improving AI systems?
4. How can alignment help make AI systems safer and more reliable?
5. Why is it important for society to guide the development of AI technologies responsibly?

Reflection Activity

Write a short paragraph explaining why powerful technologies require ethical guidance. You may compare artificial intelligence to other technologies such as medicine, transportation, or engineering, where safety and responsibility are important.

Worksheet

Chapter 11 – Fine-Tuning and Alignment

Part 1: Vocabulary Match

Match each term with the correct definition.

Term	Definition
Fine-Tuning	A. Improving a trained model using additional examples and adjustments
Alignment	B. Ensuring AI systems behave according to human values and safety principles
Human Feedback	C. Evaluation provided by people to help improve AI responses

Term	Definition
Safety	D. Protecting users and society from harmful outcomes

Part 2: Fill in the Blank

1. __________ is the process of refining a trained AI model to improve its performance.
2. __________ ensures that AI systems behave in ways consistent with human values.
3. Researchers often use __________ to evaluate and improve model responses.
4. AI systems must be designed with __________ to prevent harmful or misleading outputs.
5. Fine-tuning helps models produce more __________ and useful responses.

Part 3: Short Answer

1. Explain why a language model may need improvement after its initial training.
2. What role does human feedback play in refining AI systems?
3. Why is alignment important when developing AI technologies?
4. How can fine-tuning make AI systems more useful in education?

Part 4: Critical Thinking

Imagine you are part of a team developing an AI assistant for students.

Write 4–5 sentences explaining how fine-tuning and alignment would help ensure the system provides safe, accurate, and helpful educational support.

Quiz

Chapter 11 – Fine-Tuning and Alignment

Multiple Choice

1. What is fine-tuning in artificial intelligence?
   A. Building computer hardware
   B. Improving a trained model using additional adjustments and examples
   C. Deleting training data
   D. Programming websites

2. What does alignment mean in AI systems?
   A. Organizing computer files
   B. Ensuring AI behavior follows human values and safety standards
   C. Increasing internet speed
   D. Changing computer screens

3. Why is human feedback used in AI development?
   A. To repair computers
   B. To evaluate and improve model responses
   C. To build datasets
   D. To control internet traffic

4. Why are safety guidelines important in AI systems?
   A. They reduce the cost of computers
   B. They help prevent harmful or misleading outputs
   C. They increase computer memory
   D. They simplify programming languages

5. What is the goal of alignment in AI systems?
   A. Making AI systems faster
   B. Ensuring AI systems behave responsibly and helpfully
   C. Reducing computer storage
   D. Eliminating training data

Short Answer

6. Why is fine-tuning important for improving language models?
7. How does alignment help ensure AI systems benefit society?

Answer Key

Multiple Choice

1. **B** — Improving a trained model using additional adjustments and examples
2. **B** — Ensuring AI behavior follows human values and safety standards
3. **B** — To evaluate and improve model responses

4. **B** — They help prevent harmful or misleading outputs
5. **B** — Ensuring AI systems behave responsibly and helpfully

Fill in the Blank

1. **Fine-tuning**
2. **Alignment**
3. **human feedback**
4. **safety**
5. **accurate**

Short Answer (Example Responses)

6. Fine-tuning improves a language model by adjusting it with additional examples and feedback, helping it produce more accurate and useful responses.
7. Alignment ensures that AI systems follow ethical guidelines and human values, reducing harmful outputs and improving reliability.

Chapter 12 — Prompting: How Humans Communicate with Language Models

Learning Objectives

By the end of this chapter, students will be able to:

1. Explain what a prompt is in artificial intelligence systems.
2. Describe how prompts guide the responses generated by language models.
3. Understand why clear prompts produce better results from AI systems.
4. Recognize how prompting allows humans to collaborate with AI tools.
5. Identify basic strategies for writing effective prompts.

12.1 What Is a Prompt?

In artificial intelligence systems, a prompt is the input or instruction that a user provides to a language model. The prompt tells the AI system what information is being requested or what task needs to be performed.

Prompts can take many forms. A prompt may be a question, a statement, a request for explanation, or a command to perform a specific task. For example, a user might ask an AI system to summarize a paragraph, explain a scientific concept, generate ideas for a project, or create a lesson plan.

The quality of the prompt plays a major role in determining the quality of the response. A clear and detailed prompt helps the model understand the user's goal, while a vague prompt may produce responses that are incomplete, inaccurate, or confusing.

In this way, prompts function as the bridge between human intention and machine-generated responses.

As artificial intelligence becomes more common in education, business, and everyday life, the ability to communicate effectively with AI is becoming an important digital literacy skill. Just as students learn how to ask effective questions in a classroom, they must also learn how to provide clear instructions to intelligent systems.

The study and practice of communicating effectively with artificial intelligence is commonly known as prompt engineering. Readers interested in exploring this topic further may refer to *What to Say When You Talk to AI: Prompt Engineering Guide* by Susie Hala, which provides practical frameworks, real-world examples, and step-by-step techniques for improving human–AI communication.

12.2 Communication Between Humans and Machines

Human communication with AI systems depends on written instructions rather than voice tone or facial expressions. Language models interpret prompts by analyzing the words provided and comparing them with patterns learned during training.

When a user writes a prompt, the model examines the sequence of tokens within the prompt and predicts the most appropriate response based on its parameters and contextual understanding.

Although AI systems can generate sophisticated responses, they rely entirely on the information provided in the prompt and the patterns learned during training. For this reason, effective communication with AI systems requires thoughtful input from the user.

The interaction between prompts and responses creates a collaborative process in which humans guide the system while the model generates information based on its learned patterns.

12.3 The Importance of Clear Instructions

A well-written prompt often includes clear instructions about what type of response is desired. For example, a prompt may specify the level of detail, the subject area, or the format of the answer.

Consider the difference between the following prompts:

“Explain gravity.”

and

“Explain gravity in simple terms suitable for a middle school science class.”

The second prompt provides additional context that helps the AI system generate a more useful and appropriate explanation.

Clear prompts reduce ambiguity and help the model produce responses that align more closely with the user’s needs.

12.4 Prompt Design and Prompt Engineering

As AI systems become more widely used, researchers and professionals have begun studying the science of writing effective prompts. This practice is known as **prompt engineering**.

Prompt engineering involves designing prompts in ways that guide AI systems toward accurate, structured, and relevant responses. By adjusting the wording, structure, and level of detail in a prompt, users can influence how the model interprets the request.

For example, a user might ask a model to generate a list, write an explanation, summarize a document, or provide step-by-step instructions. Each prompt structure encourages a different type of response.

Understanding prompt design helps users interact more effectively with AI systems and obtain more useful results.

12.5 Prompts in Education

In classrooms, prompting allows students and teachers to use AI as a learning assistant.

Students can ask questions about academic subjects, request explanations of difficult concepts, or explore new ideas through guided prompts. Teachers may use prompts to generate practice questions, summarize lessons, or develop educational materials.

When used responsibly, prompting can support curiosity and encourage deeper learning. However, it is important for students to verify information and use AI as a tool for exploration rather than a replacement for critical thinking.

Learning how to write effective prompts helps students develop stronger communication skills and better understand how AI systems operate.

12.6 Prompts and Creativity

Prompting also enables creative collaboration between humans and AI systems.

Writers may use prompts to generate story ideas or expand creative writing projects. Designers might ask AI systems for inspiration when brainstorming visual concepts. Researchers may use prompts to explore new perspectives on complex topics.

In each case, the human provides the direction while the AI system generates possible responses based on patterns learned during training.

This collaborative process demonstrates that AI systems function best when guided by human curiosity and imagination.

12.7 Prompting in Modern AI Systems

Many widely used AI tools rely on prompts to interact with users. Systems such as ChatGPT, Claude, and Gemini generate responses by analyzing the prompts provided by users.

These systems demonstrate how prompting allows humans to communicate with machines using natural language. As AI technology continues to evolve, the ability to design clear and effective prompts will become an increasingly valuable skill.

Understanding prompting helps students recognize that artificial intelligence systems depend not only on advanced algorithms but also on thoughtful human input.

Chapter Summary

A prompt is the input or instruction that a user provides to a language model. Prompts guide the responses generated by AI systems and serve as the primary method through which humans communicate with artificial intelligence.

Clear and detailed prompts help language models generate more accurate and relevant responses. Prompt engineering is the practice of designing prompts in ways that guide AI systems toward useful results.

Prompting allows humans to collaborate with AI tools in education, research, and creative work. By learning how to communicate effectively with language models, students can use AI responsibly while developing stronger problem-solving and communication skills.

In the next chapter, we will explore **limitations and challenges of large language models**, examining why AI systems sometimes produce incorrect or misleading responses and how researchers work to improve them.

Reflection Section

Chapter 12 – Prompting: How Humans Communicate with Language Models

Reflection helps students think about how communication influences the results produced by artificial intelligence systems.

1. Why is a prompt necessary when interacting with a language model?
2. How can a clear prompt improve the quality of an AI response?
3. Why might a vague or unclear prompt lead to confusing results?
4. How can prompting help students learn more effectively when using AI tools?
5. Why is it important to combine AI assistance with human critical thinking?

Reflection Activity

Write a short paragraph explaining how asking better questions can improve learning. Describe a situation where a detailed question would help a teacher or AI system provide a clearer explanation.

Worksheet

Chapter 12 – Prompting

Part 1: Vocabulary Match

Match the term with the correct definition.

Term	Definition
Prompt	A. The input or instruction given to an AI system
Prompt Engineering	B. The practice of designing prompts to guide AI responses
Response	C. The output generated by an AI system
Interaction	D. Communication between humans and AI systems

Part 2: Fill in the Blank

1. A __________ is the instruction or question given to a language model.
2. The answer generated by an AI system is called a __________.
3. Writing effective prompts is known as __________.
4. Clear prompts help reduce __________ in AI responses.
5. AI systems analyze the __________ provided by the user to generate responses.

Part 3: Short Answer

1. Explain what a prompt is in artificial intelligence systems.
2. Why are clear prompts important when interacting with AI tools?
3. What is prompt engineering?
4. How can prompting help students learn new information?

Part 4: Critical Thinking

Imagine you are using an AI system to help with a history assignment.

Write 4–5 sentences describing how you would write a detailed prompt to receive a clear explanation about an important historical event.

Quiz

Chapter 12 – Prompting

Multiple Choice

1. What is a prompt in artificial intelligence systems?
   A. A computer program
   B. The instruction or question given to an AI model
   C. A dataset used for training
   D. A computer network

2. Why are clear prompts important?
   A. They increase internet speed
   B. They help the AI system understand the user's request
   C. They reduce computer memory usage
   D. They eliminate datasets

3. What is prompt engineering?
   A. Designing computer hardware
   B. Creating training datasets
   C. Writing prompts that guide AI systems toward better responses
   D. Programming operating systems

4. What happens if a prompt is unclear or vague?
   A. The AI system shuts down
   B. The response may be confusing or incomplete
   C. The dataset is deleted
   D. The computer stops working

5. How do humans collaborate with AI systems?
   A. By writing prompts and interpreting responses
   B. By replacing datasets
   C. By building computer chips
   D. By changing programming languages

Short Answer

6. Give one example of a good prompt that could help a student learn a scientific concept.
7. Why should students verify information generated by AI systems?

Answer Key

Multiple Choice

1. **B** — The instruction or question given to an AI model
2. **B** — They help the AI system understand the user's request
3. **C** — Writing prompts that guide AI systems toward better responses
4. **B** — The response may be confusing or incomplete
5. **A** — By writing prompts and interpreting responses

Fill in the Blank

1. **prompt**
2. **response**
3. **prompt engineering**
4. **ambiguity**
5. **prompt**

Short Answer (Example Responses)

6. Example: "Explain the process of photosynthesis in simple terms suitable for a middle school student."
7. AI systems sometimes generate incorrect or incomplete information, so students should verify answers using reliable sources and critical thinking.

Chapter 13 — Limitations of Large Language Models: Why AI Is Powerful but Not Perfect

Learning Objectives

By the end of this chapter, students will be able to:

1. Explain why large language models have limitations.
2. Identify common types of errors that language models may produce.
3. Understand why AI systems sometimes generate incorrect information.
4. Recognize the importance of human oversight when using AI tools.
5. Develop critical thinking skills when evaluating AI-generated responses.

13.1 Intelligence Without Understanding

Large language models can generate detailed responses that appear intelligent. However, it is important to recognize that these systems do not truly understand the world the way humans do.

Humans learn through experiences, emotions, and interactions with their environment. They connect knowledge with personal understanding and real-world observations.

Language models operate differently. They analyze patterns in text and generate responses based on probabilities learned during training. Although the responses may appear thoughtful, they are produced through mathematical calculations rather than personal understanding.

Because of this difference, language models sometimes produce answers that sound convincing but are incorrect.

13.2 The Problem of Hallucinations

One of the most well-known limitations of language models is a phenomenon often called **AI hallucination**.

An AI hallucination occurs when a language model generates information that appears believable but is actually inaccurate or fabricated. This can happen because the model is attempting to produce a response based on patterns in language rather than verified facts.

For example, a model might generate a fictional source, an incorrect date, or an explanation that sounds logical but is not supported by reliable evidence.

Researchers continue developing methods to reduce hallucinations, but it remains important for users to verify information generated by AI systems.

13.3 Bias in Training Data

Another challenge in artificial intelligence is **bias**.

Language models learn from large collections of text, and these datasets may contain cultural biases, stereotypes, or incomplete perspectives. When a model learns patterns from such data, it may unintentionally reproduce those patterns in its responses.

Developers work to reduce bias by improving datasets and refining training methods. However, because language models learn from human-created information, completely eliminating bias is extremely difficult.

Understanding bias helps students recognize why responsible AI development requires continuous evaluation and improvement.

13.4 Dependence on Training Data

The knowledge of a language model depends heavily on the data used during training.

If important information is missing from the training data, the model may struggle to answer certain questions accurately. In addition, models typically do not update their knowledge continuously unless they are retrained or connected to external information systems.

Because of this limitation, language models may not always have the most current information about recent events or discoveries.

Students should understand that AI-generated responses represent patterns learned from past data rather than a complete or constantly updated knowledge base.

13.5 Misinterpretation of Context

Even though language models analyze context to interpret meaning, they can sometimes misunderstand complex or ambiguous instructions.

A prompt that lacks clarity may lead the system to generate an answer that does not match the user's intention. In other cases, subtle wording differences can change how the model interprets the request.

This limitation highlights the importance of clear communication when interacting with AI systems.

Users often improve results by rewriting prompts, adding additional details, or asking follow-up questions.

13.6 Overreliance on AI

As AI tools become more common, there is a risk that people may rely too heavily on automated systems.

While language models can provide helpful information, they should not replace independent thinking or careful research. Students and professionals must continue developing their own knowledge and analytical skills.

AI tools are most effective when used as assistants that support learning rather than as substitutes for human reasoning.

Responsible users treat AI-generated responses as starting points for further investigation rather than final answers.

13.7 Human Oversight

Because of the limitations described in this chapter, human oversight remains essential in the use of artificial intelligence.

Teachers, researchers, and professionals must evaluate the information produced by AI systems and verify its accuracy before relying on it.

Human judgment provides something that AI systems cannot replicate: ethical reasoning, real-world experience, and the ability to evaluate information critically.

By combining AI capabilities with human wisdom, society can use artificial intelligence as a powerful tool while minimizing potential risks.

13.8 Learning to Use AI Responsibly

The purpose of studying AI limitations is not to discourage the use of technology but to encourage responsible and informed use.

Students who understand both the strengths and weaknesses of AI systems are better prepared to use these tools effectively.

Responsible AI users ask thoughtful questions, verify important information, and consider the ethical implications of technology.

In the future, many careers will involve collaboration between humans and intelligent machines. Developing critical thinking skills will help students navigate this evolving relationship with confidence.

Chapter Summary

Large language models are powerful tools for generating text, answering questions, and assisting with many types of tasks. However, these systems have important limitations.

Because language models rely on statistical patterns rather than true understanding, they may sometimes produce incorrect or misleading information. Problems such as hallucinations, bias in training data, and limitations in contextual interpretation can affect the accuracy of AI-generated responses.

The knowledge of a language model is also limited by the data used during training, meaning it may not always have access to the most current information.

For these reasons, human oversight and critical thinking remain essential when using AI systems. By understanding these limitations, students can learn to use artificial intelligence responsibly while continuing to develop their own knowledge and reasoning skills.

Chapter 13 Summary

Limitations of Large Language Models: Why AI Is Powerful but Not Perfect

Large language models are powerful tools that can generate text, answer questions, and assist with many types of tasks. However, these systems also have important limitations. Unlike humans, AI systems do not truly understand the world. Instead, they analyze patterns in data and generate responses based on probabilities learned during training.

One common limitation is the possibility of **AI hallucinations**, where a model produces information that appears convincing but is actually incorrect or fabricated. Another challenge is **bias**, which can occur when models learn patterns from datasets that contain incomplete or biased information.

Language models are also limited by the data used during training. If certain information is missing or outdated, the system may not provide accurate answers. In addition, models can sometimes misunderstand context or respond incorrectly if prompts are unclear.

Because of these limitations, human oversight remains essential when using AI systems. Students and professionals must evaluate AI-generated responses carefully and verify important information. Understanding these limitations helps users apply AI responsibly while continuing to develop critical thinking skills.

Reflection Section

Chapter 13 – Limitations of Large Language Models

Reflection questions help students think critically about the strengths and weaknesses of AI systems.

1. Why do large language models sometimes produce incorrect information?
2. What is an AI hallucination and why does it occur?
3. How can bias in training data influence the responses generated by AI systems?
4. Why is it important for users to verify information produced by AI tools?
5. How can critical thinking help people use AI more responsibly?

Reflection Activity

Write a short paragraph explaining why technology should always be used with careful judgment. Use examples from everyday life or education to describe situations where verifying information is important.

Worksheet

Chapter 13 – Limitations of Large Language Models

Part 1: Vocabulary Match

Match each term with the correct definition.

Term	Definition
AI Hallucination	A. Incorrect or fabricated information generated by an AI system
Bias	B. Patterns learned from data that may reflect unfair or incomplete perspectives
Training Data	C. Information used to teach a machine learning model
Human Oversight	D. Evaluation and supervision by people to ensure accuracy and responsibility

Part 2: Fill in the Blank

1. An __________ occurs when an AI system generates incorrect or fabricated information.
2. __________ refers to unfair patterns that may appear in data used to train AI systems.
3. Language models learn patterns from __________ used during training.
4. __________ helps ensure that AI-generated responses are accurate and responsible.
5. Users should apply __________ thinking when evaluating AI responses.

Part 3: Short Answer

1. Explain why large language models do not truly understand the world like humans do.
2. What is one example of a limitation of AI systems?
3. Why can bias occur in AI models?
4. Why should students verify information generated by AI tools?

Part 4: Critical Thinking

Imagine a student is using an AI system to research an important historical event.

Write 4–5 sentences explaining why the student should verify the information generated by the AI and consult additional sources.

Quiz

Chapter 13 – Limitations of Large Language Models

Multiple Choice

1. Why do large language models sometimes produce incorrect answers?
   A. They rely on statistical patterns rather than real understanding
   B. They cannot process language
   C. They cannot store information
   D. They cannot generate sentences

2. What is an AI hallucination?
   A. A visual display on a computer screen
   B. A fictional or incorrect answer generated by an AI system
   C. A computer virus
   D. A programming language

3. What can cause bias in AI systems?
   A. The color of the computer screen
   B. Patterns learned from biased training data
   C. The speed of the internet
   D. The size of the keyboard

4. Why is human oversight important when using AI systems?
   A. Humans must evaluate and verify information
   B. Humans control internet connections
   C. Humans design computer screens
   D. Humans build keyboards

5. What skill helps users evaluate AI-generated information effectively?
   A. Memorization
   B. Critical thinking
   C. Typing speed
   D. Computer repair

Short Answer

6. What is one example of a limitation of large language models?
7. Why is it important to combine AI tools with human judgment?

Answer Key

Multiple Choice

1. **A** — They rely on statistical patterns rather than real understanding
2. **B** — A fictional or incorrect answer generated by an AI system
3. **B** — Patterns learned from biased training data
4. **A** — Humans must evaluate and verify information
5. **B** — Critical thinking

Fill in the Blank

1. **AI hallucination**
2. **Bias**
3. **training data**
4. **Human oversight**
5. **critical**

Short Answer (Example Responses)

6. One limitation of language models is that they may generate incorrect or misleading information because they rely on patterns rather than true understanding.
7. Human judgment helps verify information, evaluate accuracy, and ensure that AI tools are used responsibly.

Chapter 14 — Human–AI Collaboration: Working Together with Intelligent Machines

Learning Objectives

By the end of this chapter, students will be able to:

1. Explain how humans and AI systems collaborate to solve problems.
2. Identify areas where human intelligence and machine intelligence complement each other.

3. Understand the advantages of combining human creativity with machine processing power.
4. Recognize the role of human judgment in guiding AI technologies.
5. Explore how collaboration between humans and AI may shape future careers and industries.

14.1 The Idea of Collaborative Intelligence

Artificial intelligence is often described as a technology that performs tasks traditionally associated with human thinking. However, the most powerful use of AI is not replacing people but working alongside them.

This partnership is known as **collaborative intelligence**—a system in which humans and machines contribute different strengths to accomplish shared goals.

Humans bring creativity, ethical judgment, emotional awareness, and real-world experience. Machines contribute speed, data processing, pattern recognition, and the ability to analyze large amounts of information.

When these abilities are combined, they create a partnership that allows complex problems to be solved more effectively than either humans or machines could manage alone.

14.2 Human Strengths

Human intelligence includes abilities that remain difficult for machines to replicate.

People can understand emotions, interpret social cues, and make decisions that consider ethical and cultural factors. Humans also demonstrate creativity by imagining ideas that have never existed before.

Another important strength of human intelligence is the ability to adapt to unfamiliar situations. Humans can apply knowledge from different experiences and modify their reasoning when circumstances change.

These capabilities allow humans to guide the direction and purpose of technology.

14.3 Machine Strengths

Artificial intelligence systems excel at tasks that involve large amounts of data and repetitive analysis.

Language models can examine enormous collections of text and identify patterns that would be difficult for humans to detect manually. AI systems can also perform calculations at extremely high speeds and analyze information continuously without fatigue.

These abilities allow machines to support decision-making by providing insights drawn from large datasets.

However, machines do not possess emotions, personal experiences, or moral judgment. This is why human supervision remains essential in guiding how AI systems are used.

14.4 Collaboration in Education

One of the most promising areas of human–AI collaboration is education.

AI systems can help students explore complex topics, generate explanations, and provide practice exercises tailored to individual learning needs. Teachers can use AI tools to organize materials, design lessons, and analyze learning progress.

In this partnership, the teacher provides mentorship, empathy, and guidance, while AI systems support the learning process through information and analysis.

This combination can help create more personalized and accessible educational experiences for students.

14.5 Collaboration in Science and Medicine

Human–AI collaboration is also transforming scientific research and healthcare.

In medical research, AI systems analyze large datasets from clinical studies, helping scientists identify patterns that could lead to new treatments. Doctors can use AI tools to examine medical images and detect early signs of disease.

Despite these technological advances, medical professionals remain responsible for interpreting results and making final decisions about patient care.

The combination of machine analysis and human expertise can improve diagnostic accuracy and accelerate scientific discovery.

14.6 Creativity and Innovation

AI systems are increasingly being used as tools for creative exploration.

Artists, writers, musicians, and designers sometimes use AI-generated suggestions as inspiration for new projects. Rather than replacing human creativity, these systems provide new perspectives that help creators explore different possibilities.

This type of collaboration demonstrates that technology can expand the creative process by introducing ideas that humans may refine or reinterpret.

The final creative work still reflects human intention and imagination.

14.7 Human Responsibility

While AI can provide powerful tools for analysis and creativity, responsibility for how these tools are used remains with humans.

People must decide how AI systems are applied in society, how they are regulated, and how their benefits are shared fairly.

Human oversight ensures that technology supports ethical principles such as fairness, transparency, and respect for human rights.

Students studying artificial intelligence should understand that technological progress must always be guided by thoughtful decision-making.

14.8 Collaboration in Modern AI Systems

Many modern AI tools demonstrate the idea of human–AI collaboration. Systems such as ChatGPT, Claude, and Gemini allow users to interact with language models that assist with writing, research, and learning.

In these systems, humans provide prompts, direction, and judgment, while AI systems generate information based on patterns learned during training.

This partnership illustrates how humans and intelligent machines can work together to solve problems and explore ideas.

Chapter Summary

Human–AI collaboration represents a partnership between human intelligence and machine intelligence. Humans contribute creativity, ethical reasoning, emotional awareness, and real-world experience, while AI systems provide rapid data analysis and pattern recognition.

By combining these strengths, humans and machines can work together to solve complex problems in fields such as education, science, medicine, and creative industries.

Despite the capabilities of AI systems, human oversight remains essential. People must guide how AI technologies are used and ensure that they support ethical values and societal well-being.

Understanding the relationship between human intelligence and artificial intelligence prepares students to participate responsibly in a future where collaboration between humans and machines becomes increasingly common.

Reflection Section

Chapter 14 – Human–AI Collaboration: Working Together with Intelligent Machines

Reflection questions help students think about how humans and artificial intelligence can work together effectively.

1. What does the term **human–AI collaboration** mean?
2. What strengths do humans bring to partnerships with AI systems?
3. What tasks are AI systems particularly good at performing?
4. Why is human oversight important when using AI technologies?
5. How might human–AI collaboration change future careers and workplaces?

Reflection Activity

Write a short paragraph describing a situation where humans and machines could work together to solve a problem. Explain what role the human would play and what role the AI system would perform.

Worksheet

Chapter 14 – Human–AI Collaboration

Part 1: Vocabulary Match

Match each term with the correct definition.

Term	Definition
Collaboration	A. Working together to accomplish a shared goal
Artificial Intelligence	B. Computer systems designed to perform tasks that require intelligence

Term	Definition
Human Judgment	C. The ability of people to evaluate situations using reasoning and experience
Data Analysis	D. The process of examining information to identify patterns or insights

Part 2: Fill in the Blank

1. __________ refers to humans and machines working together to solve problems.
2. Humans contribute creativity, empathy, and __________ when collaborating with AI.
3. AI systems are especially good at analyzing large amounts of __________.
4. __________ ensures that AI systems are used responsibly and ethically.
5. Collaboration between humans and AI can lead to new __________ and innovations.

Part 3: Short Answer

1. Explain how human intelligence and artificial intelligence can complement each other.
2. Why are AI systems useful when working with large datasets?
3. What role do humans play in guiding AI technologies?
4. How might AI tools support teachers and students in the classroom?

Part 4: Critical Thinking

Imagine a team of scientists using AI tools to study climate change.

Write 4–5 sentences explaining how scientists and AI systems could collaborate to analyze environmental data and develop solutions.

Quiz

Chapter 14 – Human–AI Collaboration

Multiple Choice

1. What does human–AI collaboration mean?
   A. Replacing humans with machines
   B. Humans and AI systems working together to solve problems

C. Eliminating technology from workplaces
D. Using computers only for entertainment

2. Which of the following is a strength of human intelligence?
   A. Rapid mathematical calculations
   B. Emotional understanding and ethical reasoning
   C. Processing billions of data points per second
   D. Storing large datasets

3. What is a strength of AI systems?
   A. Emotional awareness
   B. Creative imagination
   C. Analyzing large amounts of data quickly
   D. Personal life experiences

4. Why is human oversight important when using AI?
   A. Humans must guide technology responsibly
   B. Humans control internet connections
   C. Humans operate keyboards
   D. Humans repair computer screens

5. How can AI support education?
   A. By replacing teachers
   B. By assisting with explanations, practice exercises, and learning tools
   C. By removing textbooks
   D. By eliminating homework

Short Answer

Give one example of how humans and AI systems can collaborate in science or medicine.

6. Why is human creativity still important in an age of artificial intelligence?

Answer Key

Multiple Choice

1. **B** — Humans and AI systems working together to solve problems
2. **B** — Emotional understanding and ethical reasoning
3. **C** — Analyzing large amounts of data quickly
4. **A** — Humans must guide technology responsibly
5. **B** — By assisting with explanations, practice exercises, and learning tools

Fill in the Blank

1. **Collaboration**
2. **judgment**
3. **data**
4. **Human oversight**
5. **discoveries**

Short Answer (Example Responses)

6. Scientists may use AI to analyze medical images or climate data, while humans interpret the results and make decisions about treatments or environmental policies.
7. Human creativity is important because people imagine new ideas, design solutions, and guide technology in ways that machines alone cannot achieve.

Chapter 15 — The Future of Large Language Models: The Next Stage of Intelligent Systems

Learning Objectives

By the end of this chapter, students will be able to:

1. Describe how large language models may evolve in the future.
2. Understand the concept of multimodal artificial intelligence systems.
3. Explain how AI assistants may become more integrated into everyday life.
4. Identify emerging technologies related to language models and intelligent agents.
5. Reflect on how future AI systems may influence education, work, and society.

15.1 A Rapidly Evolving Technology

Artificial intelligence is advancing at a remarkable pace. Only a few years ago, language models were capable of completing short sentences or answering basic questions. Today, they can generate essays, translate languages, summarize research papers, and assist with complex tasks.

Large language models continue to evolve as researchers develop new training methods, improved architectures, and more powerful computing systems. Each new generation of models becomes better at understanding context, producing detailed explanations, and assisting users with a wide range of activities.

The future of language models will likely involve systems that are more accurate, more efficient, and capable of interacting with many forms of information beyond text.

15.2 Multimodal Intelligence

One of the most important developments in artificial intelligence is the rise of **multimodal AI systems**.

Multimodal systems are capable of processing multiple types of information at the same time. Instead of analyzing only text, these systems may also interpret images, audio, video, and other forms of data.

For example, a multimodal AI system might analyze a photograph and generate a written explanation of what it sees. It could also interpret spoken language and respond with both text and voice.

By combining different types of information, multimodal AI systems can create richer and more natural interactions between humans and machines.

15.3 AI Assistants in Everyday Life

Large language models are increasingly becoming the foundation for digital assistants that support everyday activities.

In the future, AI assistants may help people manage schedules, organize information, conduct research, and assist with learning new skills. Students may interact with AI tutors that provide personalized explanations and learning support.

Professionals may use AI assistants to analyze data, draft documents, or explore creative ideas. These systems may also assist with translation and communication across languages.

As AI assistants become more capable, they may function as collaborative partners that help people complete tasks more efficiently.

15.4 Intelligent Agents

Another emerging concept in artificial intelligence is the development of **intelligent agents**.

An intelligent agent is a system designed to perform tasks autonomously based on goals provided by a user. Instead of responding to a single prompt, an AI agent can plan a sequence of actions to accomplish a larger objective.

For example, a research agent might gather information from multiple sources, organize the findings, and produce a summary report. A scheduling agent might coordinate meetings, send reminders, and adjust plans based on changing priorities.

These systems represent a shift from simple question–answer interactions toward AI systems that can perform complex tasks with greater independence.

15.5 The Role of Human Guidance

Despite these technological advances, human guidance will remain essential in shaping the future of artificial intelligence.

Developers, educators, policymakers, and researchers must ensure that AI systems are designed responsibly and used in ways that benefit society.

Human values, ethical principles, and cultural perspectives must guide how AI technologies are developed and applied.

Students learning about AI today may become the scientists, engineers, and leaders who determine how these technologies evolve in the future.

15.6 AI in Education and Learning

The future of AI may transform how people learn and share knowledge.

Language models can help create personalized learning experiences that adapt to each student's pace and interests. AI tutors may provide explanations, generate practice exercises, and help students explore complex topics.

Teachers may use AI tools to design lessons, analyze student progress, and develop new educational materials.

When used responsibly, these technologies can help make education more accessible and engaging for learners around the world.

15.7 Opportunities and Challenges

The future of large language models brings both exciting opportunities and important challenges.

AI systems may help accelerate scientific discoveries, improve healthcare research, and support creative innovation. At the same time, society must address questions related to privacy, fairness, reliability, and the responsible use of technology.

Balancing innovation with responsibility will be one of the most important challenges of the coming decades.

Students who understand both the benefits and risks of AI will be better prepared to contribute to thoughtful decision-making about these technologies.

15.8 The Expanding AI Ecosystem

Many advanced AI tools already demonstrate the expanding capabilities of language models. Systems such as ChatGPT, Claude, and Gemini illustrate how AI assistants can support communication, research, and learning.

As technology continues to evolve, these systems may become even more capable and integrated into everyday life.

Understanding the future direction of AI helps students prepare for a world where collaboration between humans and intelligent machines becomes increasingly common.

Chapter 15 Summary

The Future of Large Language Models

Large language models are evolving rapidly as artificial intelligence technology continues to advance. Early AI systems were limited in their ability to understand and generate language, but modern language models can perform complex tasks such as answering questions, summarizing documents, and assisting with research and writing.

One important development in AI is the growth of **multimodal systems**, which can process multiple forms of information such as text, images, audio, and video. These systems allow AI to interact with humans in more natural and flexible ways.

Another emerging concept is the use of **AI assistants and intelligent agents**. These systems can help users organize information, manage tasks, and perform research more efficiently. Instead of responding to single questions, intelligent agents may complete entire sequences of tasks.

Although the future of AI offers many opportunities, it also presents challenges related to responsibility, fairness, and ethical use. Human guidance will remain essential to ensure that AI technologies support society in positive and meaningful ways.

Understanding the future of large language models helps students prepare for a world where humans and intelligent machines collaborate to solve complex problems.

Reflection Section

Chapter 15 – The Future of Large Language Models

Reflection questions encourage students to think about how artificial intelligence may influence society in the future.

1. How have large language models improved in recent years?
2. What does the term **multimodal AI** mean?
3. How might AI assistants support people in everyday activities?
4. Why is human guidance important when developing advanced AI systems?
5. What opportunities and challenges might future AI technologies create?

Reflection Activity

Write a short paragraph describing how AI might change the way people learn or work in the future. Consider both the potential benefits and the responsibilities that come with new technologies.

Worksheet

Chapter 15 – The Future of Large Language Models

Part 1: Vocabulary Match

Match each term with the correct definition.

Term	Definition
Multimodal AI	A. AI systems that can process multiple types of information
AI Assistant	B. A digital system that helps users perform tasks or answer questions
Intelligent Agent	C. A system that can perform tasks autonomously to achieve goals
Innovation	D. The creation of new ideas, technologies, or methods

Part 2: Fill in the Blank

1. __________ AI systems can process information such as text, images, and audio.
2. An __________ can help users perform tasks such as organizing schedules or answering questions.
3. An __________ is a system that can perform tasks automatically to achieve a goal.
4. The future of AI will involve both new opportunities and new __________.
5. Human __________ is necessary to ensure that AI systems are used responsibly.

Part 3: Short Answer

1. Explain what multimodal AI systems are.
2. How might AI assistants help students in school?
3. What is an intelligent agent and how does it differ from a simple AI response system?
4. Why is responsible development important in the future of artificial intelligence?

Part 4: Critical Thinking

Imagine that AI systems become powerful assistants in many professions.

Write 4–5 sentences explaining how AI might help scientists, teachers, or engineers solve complex problems in the future.

Quiz

Chapter 15 – The Future of Large Language Models

Multiple Choice

1. What does the term **multimodal AI** mean?
   A. AI systems that only process text

B. AI systems that process multiple types of information such as text and images
C. AI systems used only in computers
D. AI systems that only perform calculations

2. What is an AI assistant?
   A. A robot designed only for factories
   B. A system that helps users perform tasks or answer questions
   C. A computer chip
   D. A programming language

3. What is an intelligent agent?
   A. A human programmer
   B. A system that can perform tasks autonomously to achieve goals
   C. A computer monitor
   D. A data storage device

4. Why is human guidance important in the future of AI?
   A. Humans design computer keyboards
   B. Humans must ensure AI systems are used responsibly
   C. Humans control internet signals
   D. Humans store computer memory

5. What is one potential benefit of future AI systems?
   A. Eliminating all human work
   B. Supporting research, learning, and problem solving
   C. Removing the need for education
   D. Replacing human creativity

Short Answer

6. How might AI systems help improve education in the future?
7. Why must society consider ethical questions when developing AI technologies?

Answer Key

Multiple Choice

1. **B** — AI systems that process multiple types of information such as text and images
2. **B** — A system that helps users perform tasks or answer questions
3. **B** — A system that can perform tasks autonomously to achieve goals
4. **B** — Humans must ensure AI systems are used responsibly
5. **B** — Supporting research, learning, and problem solving

Chapter 16 — Responsible AI: Building a Future Where Humans and Intelligent Machines Thrive Together

Learning Objectives

By the end of this chapter, students will be able to:

1. Explain what responsible AI means.
2. Describe why ethical guidelines are necessary when developing artificial intelligence systems.
3. Identify key principles that guide responsible AI development.
4. Understand the role of individuals, institutions, and governments in shaping AI policy.
5. Reflect on how future generations can use AI technologies wisely and ethically.

16.1 The Responsibility of Intelligence

Throughout history, every major technological breakthrough has created both opportunities and responsibilities. Electricity transformed industries and homes. The internet reshaped communication and access to knowledge. Artificial intelligence represents another powerful step in humanity's technological evolution.

Large language models and other AI systems can analyze information, assist with research, and support decision-making in ways that were once impossible. However, the same power that makes AI useful also requires careful oversight.

Responsible AI means ensuring that artificial intelligence systems are designed, developed, and used in ways that benefit individuals, communities, and society as a whole.

16.2 What Is Responsible AI?

Responsible AI refers to the ethical development and use of artificial intelligence technologies. It involves designing systems that are safe, fair, transparent, and aligned with human values.

Because AI systems learn patterns from data and influence real-world decisions, developers must carefully consider how these systems affect people. Responsible AI practices help prevent harmful outcomes and promote trust in technology.

Responsible AI is not only the responsibility of engineers or researchers. Educators, policymakers, organizations, and everyday users all play a role in shaping how AI systems are used.

16.3 Principles of Ethical AI

Researchers and organizations have developed several guiding principles that help ensure AI systems are developed responsibly.

One important principle is **fairness**, which means that AI systems should treat individuals and groups equitably without reinforcing harmful biases.

Another principle is **transparency**, which refers to making AI systems understandable so that people can evaluate how decisions are made.

Accountability is also essential. When AI systems influence decisions that affect people's lives, humans must remain responsible for those outcomes.

Finally, **safety and reliability** ensure that AI systems operate consistently and minimize the risk of harmful errors.

Together, these principles help guide the ethical development of artificial intelligence.

16.4 The Role of Education

Education plays a critical role in preparing people to use artificial intelligence responsibly.

Students who understand how AI systems work are better equipped to evaluate information generated by these technologies. They can ask thoughtful questions, recognize potential limitations, and use AI tools effectively.

Teaching AI literacy also helps future professionals develop the skills needed to design responsible technologies. As artificial intelligence becomes more integrated into society, informed citizens will play an essential role in guiding its development.

By learning about AI today, students are preparing themselves to shape the future of technology tomorrow.

16.5 Human Values and Technology

Artificial intelligence does not exist in isolation. It is created by people and reflects the values, goals, and decisions of the societies that develop it.

Human values such as fairness, respect, and responsibility must guide how AI systems are used. Without thoughtful guidance, technology could be applied in ways that create unintended consequences.

Responsible AI development requires collaboration between technologists, educators, governments, and communities. Together, these groups help ensure that AI technologies serve the public good.

When technology reflects human values, it becomes a tool that supports progress rather than creating new challenges.

16.6 Global Perspectives on AI

Artificial intelligence is a global technology that affects societies around the world. Countries, organizations, and research institutions are working to establish guidelines and policies that encourage responsible AI development.

International cooperation is important because AI systems influence communication, commerce, education, and scientific research across national boundaries.

By sharing knowledge and establishing common principles, societies can work together to ensure that AI technologies are used in ways that benefit humanity.

16.7 The Role of Future Innovators

Students studying artificial intelligence today may become the researchers, engineers, policymakers, and educators who guide the future of technology.

These future innovators will help determine how AI systems are designed, how they are regulated, and how they are used in everyday life.

Understanding the principles of responsible AI prepares students to approach technological challenges with wisdom and ethical awareness.

The goal is not simply to build powerful machines, but to ensure that those machines contribute positively to human progress.

16.8 A Partnership for the Future

Artificial intelligence is not a replacement for human intelligence. Instead, it is a tool that can extend human capabilities when used responsibly.

Modern AI systems such as ChatGPT, Claude, and Gemini demonstrate how language models can support communication, learning, and research.

The future of artificial intelligence will depend on how humans choose to guide and apply these technologies.

When humans and intelligent machines work together with responsibility and ethical awareness, they can create solutions that improve lives and expand knowledge.

Chapter Summary

Responsible AI refers to the ethical development and use of artificial intelligence systems. As AI technologies become more powerful and widespread, ensuring that these systems operate safely and fairly becomes increasingly important.

Key principles of responsible AI include fairness, transparency, accountability, and reliability. These principles help guide the development of AI systems that respect human values and promote trust in technology.

Education plays a vital role in preparing individuals to use AI responsibly. Students who understand how AI systems function can evaluate their outputs critically and apply these tools effectively.

The future of artificial intelligence will be shaped by the choices people make today. By combining technological innovation with ethical responsibility, humanity can build a future in which humans and intelligent machines work together to solve complex problems and expand the boundaries of knowledge.

Chapter 16 Summary

Responsible AI: Building a Future Where Humans and Intelligent Machines Thrive Together

Responsible AI refers to the ethical development and use of artificial intelligence systems. As AI technologies become more powerful and widely used, it is important to ensure that these systems operate in ways that benefit society while minimizing risks.

Responsible AI focuses on guiding technology with human values such as fairness, transparency, accountability, and safety. These principles help ensure that AI systems are designed to treat people fairly, operate reliably, and allow humans to understand how decisions are made.

Education also plays an important role in responsible AI development. When students understand how AI works, they can evaluate information more carefully and use these technologies responsibly. Teachers, researchers, governments, and organizations must work together to create policies and practices that guide AI development.

The future of artificial intelligence will depend on how humans choose to design and use these technologies. By combining innovation with ethical responsibility, society can ensure that intelligent machines support human progress and contribute to a better future.

Reflection Section

Chapter 16 – Responsible AI

Reflection encourages students to consider how technology should be guided by ethical principles.

1. What does the term **responsible AI** mean?
2. Why are fairness and transparency important when developing AI systems?
3. Why must humans remain accountable for decisions made with AI assistance?
4. How can education help people use AI responsibly?
5. What role might future generations play in shaping how AI technologies are used?

Reflection Activity

Write a short paragraph describing how you think artificial intelligence should be used in society. Explain how people can ensure that AI systems benefit communities while protecting individuals from potential harm.

Worksheet

Chapter 16 – Responsible AI

Part 1: Vocabulary Match

Match each term with the correct definition.

Term	Definition
Responsible AI	A. Ethical development and use of artificial intelligence systems
Fairness	B. Ensuring AI systems treat people and groups equitably
Transparency	C. Making AI systems understandable so people can evaluate decisions
Accountability	D. Accepting responsibility for decisions influenced by AI systems

Part 2: Fill in the Blank

1. __________ AI refers to the ethical development and use of artificial intelligence systems.

2. __________ ensures that AI systems treat individuals and groups fairly.
3. __________ means that people can understand how AI systems make decisions.
4. Humans must remain __________ for decisions influenced by AI systems.
5. __________ helps prepare people to use AI technologies responsibly.

Part 3: Short Answer

1. Explain why responsible AI is important in modern society.
2. What are two principles that guide ethical AI development?
3. Why must humans maintain oversight of AI systems?
4. How can students contribute to responsible AI use in the future?

Part 4: Critical Thinking

Imagine that artificial intelligence becomes widely used in government decision-making.

Write 4–5 sentences explaining why transparency and accountability would be important when AI systems influence public policies.

Quiz

Chapter 16 – Responsible AI

Multiple Choice

1. What does responsible AI refer to?
   A. Faster computer hardware
   B. Ethical development and use of AI technologies
   C. Computer programming languages
   D. Internet connections

2. Why is fairness important in AI systems?
   A. It ensures computers run faster
   B. It helps ensure that AI systems treat people equitably
   C. It increases internet speed
   D. It reduces computer memory

3. What does transparency mean in artificial intelligence?
   A. Making AI systems invisible
   B. Ensuring people understand how AI systems make decisions
   C. Increasing computer storage
   D. Removing data from models

4. Why is human accountability necessary in AI system?
   A. Humans design keyboards
   B. Humans must take responsibility for decisions influenced by AI
   C. Humans build computer screens
   D. Humans manage internet signals

5. What role does education play in responsible AI development?
   A. It replaces AI systems
   B. It prepares people to understand and use AI responsibly
   C. It eliminates datasets
   D. It prevents technological progress

Short Answer

6. What is one example of how responsible AI could benefit society?
7. Why should AI technologies be guided by ethical principles?

Answer Key

Multiple Choice

1. **B** — Ethical development and use of AI technologies
2. **B** — It helps ensure that AI systems treat people equitably
3. **B** — Ensuring people understand how AI systems make decisions
4. **B** — Humans must take responsibility for decisions influenced by AI
5. **B** — It prepares people to understand and use AI responsibly

Fill in the Blank

1. **Responsible**
2. **Fairness**
3. **Transparency**
4. **accountable**
5. **Education**

Short Answer (Example Responses)

6. Responsible AI could help improve healthcare systems by analyzing medical data while ensuring patient privacy and fairness in treatment decisions.
7. Ethical principles guide AI development to ensure technology benefits society while protecting people from harmful outcomes.

GLOSSARY

Letter A

1. Access Time

The amount of time it takes for an AI system or computer to locate and retrieve data from memory or storage. Access time affects how quickly megabytes of data can be used during AI processing.

2. Activation Data

Intermediate data generated inside an AI model while it is running. Activation data is stored temporarily in memory and often measured in megabytes during model execution.

3. Adaptive Storage

A storage approach that automatically adjusts how data is stored or moved based on usage patterns. In AI systems, adaptive storage helps manage large volumes of megabytes efficiently.

4. Algorithmic Data Size

The total amount of data an algorithm needs to operate effectively. In AI, this size is often measured in megabytes or larger units, depending on model complexity.

5. Allocation (Memory Allocation)

The process of assigning a specific number of megabytes in memory or storage for an AI task, model, or dataset.

6. Analytics Dataset

A structured collection of data used for analysis or AI training. The size of an analytics dataset is commonly measured in megabytes, gigabytes, or more.

7. Annotation Storage

The space required to store labels, tags, or explanations added to AI training data. These annotations contribute additional megabytes to total dataset size.

8. Artificial Neural Network (ANN) Size

The total memory footprint of a neural network, including weights, biases, and parameters. ANN size is often described in megabytes when discussing model efficiency.

9. Asset Compression

The process of reducing the number of megabytes required to store AI-related files such as models, images, or datasets without losing essential information.

10. Available Memory

The amount of free memory, measured in megabytes, that an AI system can still use to load data, run models, or perform computations.

ANI (Artificial Narrow Intelligence)

Artificial Narrow Intelligence refers to AI systems designed to perform **one specific task or a narrow set of tasks** rather than general intelligence. ANI systems rely on data stored and processed in memory and storage measured in **megabytes, gigabytes, or more**. Examples include image recognition, speech-to-text, recommendation systems, and chatbots. ANI does not understand beyond its training scope, but it can process large amounts of data efficiently within its defined function.

Letter B

1. Bandwidth

The amount of data that can be transferred between systems in a given time. In AI systems, bandwidth determines how quickly megabytes of data can move between storage, memory, and processors.

2. Batch Processing

A method where AI systems process large groups of data all at once instead of one item at a time. Batch processing often involves handling many megabytes of data in a single operation.

3. Binary Encoding

The process of representing information using binary values (0s and 1s). All megabytes of AI data are ultimately stored and transmitted using binary encoding.

4. Bit Depth

The number of bits used to represent a single piece of data, such as a pixel or audio sample. Higher bit depth increases data quality but also increases the number of megabytes required.

5. Bitmap Data

A type of image data where each pixel is stored as a set of bits. Bitmap images can quickly grow in size, consuming large numbers of megabytes in AI vision systems.

6. Buffer Memory

Temporary memory used to hold data while it is being transferred or processed. Buffers store data in chunks measured in megabytes to keep AI systems running smoothly.

7. Bulk Data Storage

Large-scale storage designed to hold massive datasets used for AI training or analysis. Bulk storage capacity is measured in megabytes, gigabytes, or larger units.

8. Byte Stream

A continuous flow of bytes transmitted or processed by a system. AI applications often analyze byte streams that accumulate into megabytes of data.

9. Bytecode

An intermediate form of program code that is easier for machines to execute than human-written code. Bytecode files occupy storage space measured in megabytes.

10. Byzantine Fault Tolerance (BFT)

A system property that allows AI or distributed systems to continue functioning even when some components fail or act unpredictably. Implementing BFT requires additional data handling and storage overhead measured in megabytes.

Letter C

1. Cache Memory

A small, high-speed memory area that stores frequently used data so the processor can access it faster. Cache helps reduce the time needed to move megabytes of data from main memory or storage.

2. Capacity Planning

The process of estimating how much memory and storage an AI system will need. Capacity planning ensures there are enough megabytes available to handle data, models, and future growth.

3. Checkpoint File

A saved snapshot of an AI model during training. Checkpoint files are stored in memory or storage and often consume many megabytes so training can resume without starting over.

4. Chunking

The practice of breaking large datasets into smaller pieces. Chunking helps AI systems process megabytes of data more efficiently by handling them in manageable parts.

5. Cloud Storage

A method of storing data on remote servers accessed through the internet. Cloud storage manages files measured in megabytes and allows AI systems to scale beyond local hardware limits.

6. Compression

The process of reducing the size of data so it uses fewer megabytes. Compression helps AI systems store, transfer, and process data more efficiently.

7. Compute Load

The amount of processing work required to run an AI task. Compute load often increases as the number of megabytes being processed grows.

8. Context Window

The amount of data an AI model can consider at one time. Context windows consume memory measured in megabytes, especially in language and vision models.

9. Corpus

A large collection of text or data used to train AI models. A corpus is usually measured in megabytes or larger units depending on its size.

10. CUDA Memory

Specialized memory used by GPUs to accelerate AI computations. CUDA memory holds model data and intermediate results measured in megabytes during processing.

Letter D

1. Data Allocation

The process of assigning a specific amount of memory or storage for data use. In AI systems, data allocation determines how many megabytes are reserved for datasets, models, or tasks.

2. Data Buffer

A temporary storage area that holds data while it is being transferred or processed. Data buffers manage megabytes of information to keep AI systems running smoothly.

3. Data Compression

A technique used to reduce the size of data so it consumes fewer megabytes. AI systems use data compression to save storage space and speed up data transfer.

4. Data Footprint

The total amount of data an AI system uses, including datasets, models, and temporary files. A data footprint is measured in megabytes, gigabytes, or more.

5. Data Loader

A software component that loads data into memory for AI training or inference. Data loaders handle batches of data measured in megabytes.

6. Data Migration

The process of moving data from one storage system to another. During AI system upgrades, data migration may involve transferring large numbers of megabytes.

7. Data Throughput

The rate at which data is processed or transferred by a system. Higher throughput allows AI systems to move more megabytes efficiently.

8. Dataset Size

The total amount of data in a dataset used for AI training or testing. Dataset size is commonly measured in megabytes or larger units.

9. Deep Learning Model Size

The amount of memory required to store a deep learning model, including its parameters and weights. Model size is often described in megabytes to assess efficiency.

10. Disk Storage

Long-term storage used to save files, datasets, and AI models. Disk storage capacity and usage are measured in megabytes, gigabytes, or terabytes.

Letter E

1. Edge Computing

A computing approach where data is processed closer to where it is generated rather than sent to a central server. Edge computing reduces how many megabytes must be transferred over networks.

2. Embedding Size

The amount of memory required to store vector representations used by AI models. Embedding size is often measured in megabytes and affects model performance and storage needs.

3. Encoded Data

Data that has been converted into a specific format for storage or transmission. Encoded data occupies megabytes depending on the encoding method used.

4. Encoding Scheme

A rule set that defines how data is represented digitally, such as text or images. Different encoding schemes can increase or reduce the number of megabytes required.

5. Encrypted Storage

Storage that protects data by converting it into a secure format. Encryption may slightly increase the number of megabytes used due to added security information.

6. End-to-End Data Pipeline

The full path data takes from collection to processing and storage. Each stage of the pipeline manages data volumes measured in megabytes.

7. Energy Efficiency (AI Systems)

A measure of how much energy an AI system uses to process data. Systems that move and store fewer megabytes are often more energy efficient.

8. Execution Memory

The memory required while a program or AI model is running. Execution memory usage is tracked in megabytes during runtime.

9. External Storage

Storage devices located outside a computer, such as USB drives or external hard drives. External storage capacity is measured in megabytes and larger units.

10. Extracted Features

Important data patterns identified by AI models during processing. Extracted features are stored temporarily or permanently in memory measured in megabytes.

Letter F

1. Feature Map

A structured output produced by AI models, especially in image and signal processing. Feature maps are stored in memory and can consume significant megabytes during computation.

2. Feature Scaling

The process of adjusting data values to a consistent range before training an AI model. Feature scaling affects how efficiently megabytes of data are processed and stored.

3. File Allocation Table (FAT)

A file system structure that tracks where files are stored on a disk. FAT helps manage how megabytes are organized and retrieved from storage.

4. File Compression Format

A standardized way to reduce file size so it uses fewer megabytes. Compression formats help AI systems store and transfer data efficiently.

5. File System Cache

A temporary memory area that stores frequently accessed files. File system caches reduce repeated access to storage and manage megabytes more efficiently.

6. Floating-Point Data

Numerical data that includes decimal values and is commonly used in AI calculations. Floating-point data requires more bits per value, increasing memory usage measured in megabytes.

7. Frame Buffer

A region of memory that stores image or video frames before they are displayed or processed. Frame buffers often consume large numbers of megabytes in AI vision systems.

8. Fragmentation

A condition where data is stored in scattered locations instead of one continuous block. Fragmentation can reduce storage efficiency even when megabytes appear available.

9. Free Memory

The amount of unused memory available for programs or AI tasks. Free memory is measured in megabytes and determines how much additional data can be loaded.

10. Full Dataset Load

The act of loading an entire dataset into memory at once. This approach can require large amounts of megabytes and is common in AI training scenarios.

Letter G

1. Garbage Collection

An automatic memory-management process that frees memory no longer in use. Garbage collection helps reclaim megabytes so AI programs can continue running efficiently.

2. Generalization Data

Data used to test how well an AI model performs on new, unseen inputs. Storing and evaluating generalization data requires additional megabytes beyond training data.

3. Gigabyte (GB)

A larger unit of data measurement equal to 1,024 megabytes (or commonly approximated as 1,000 MB). Gigabytes are used when datasets or models grow beyond megabyte scale.

4. Graph Data Structure

A way of organizing data using nodes and connections. Graph-based AI systems store graph data in memory and storage measured in megabytes.

5. GPU Memory

Specialized memory used by graphics processing units to accelerate AI workloads. GPU memory holds model parameters and data batches measured in megabytes.

6. Gradient Data

Numerical values calculated during AI training to update model parameters. Gradient data is stored temporarily in memory and contributes to overall megabyte usage.

7. Granularity

The level of detail at which data is stored or processed. Finer granularity often increases the number of megabytes required.

8. Greedy Algorithm

An algorithm that makes the best immediate choice at each step. Greedy algorithms may reduce memory usage by limiting how many megabytes of data must be stored at once.

9. Ground Truth Data

Correct, labeled data used to train or evaluate AI systems. Ground truth datasets are stored in files measured in megabytes or larger units.

10. Growth Rate (Data)

The speed at which data size increases over time. In AI systems, a high data growth rate means storage needs measured in megabytes can expand quickly.

Letter H

1. Hard Disk Capacity

The total amount of data a hard disk can store. Capacity is measured in megabytes and larger units.

2. Hardware Acceleration

The use of specialized hardware to speed up AI processing. Acceleration often reduces how long megabytes of data stay in memory.

3. Hash Table

A data structure that stores information for fast lookup. Hash tables occupy memory measured in megabytes in large AI systems.

4. Hierarchical Storage

A storage design that uses multiple layers such as RAM, disk, and cloud. Each layer manages data measured in megabytes.

5. High-Dimensional Data

Data with many features or variables. High-dimensional datasets often require large numbers of megabytes.

6. Host Memory

Main system memory used by the CPU. Host memory stores AI data and models measured in megabytes.

7. Hybrid Storage

A combination of different storage types working together. Hybrid storage systems manage megabytes across multiple devices.

8. Hyperparameter Storage

Memory used to store configuration values for AI models. These settings are saved in files measured in megabytes.

9. Hyperscale Data

Extremely large datasets used by major AI systems. Hyperscale data is tracked from megabytes up to much larger units.

10. Heuristic Data

Information used by AI systems to guide decision-making. Heuristic data contributes to the overall data size in megabytes.

Letter I

1. Image Dataset Size

The total amount of storage required for image data. Size is measured in megabytes or more.

2. Inference Memory

The memory used when an AI model is making predictions. Inference memory usage is tracked in megabytes.

3. Input Buffer

A temporary area where incoming data is stored. Input buffers hold megabytes before processing begins.

4. Intermediate Data

Data produced during processing steps. Intermediate data can significantly increase memory usage in megabytes.

5. Index File

A file that helps locate data quickly. Index files use additional megabytes to improve performance.

6. Information Density

The amount of useful data stored within a given size. Higher density means more information per megabyte.

7. Initialization Data

Data loaded when an AI system starts. Initialization data occupies memory measured in megabytes.

8. In-Memory Processing

Processing data directly in RAM instead of storage. This approach uses large amounts of memory measured in megabytes.

9. Input Feature Size

The amount of data required to represent input features. Feature size affects how many megabytes are needed.

10. Instruction Cache

A small memory area that stores frequently used instructions. Instruction caches manage small but critical megabytes.

Letter J

1. Job Queue

A list of tasks waiting to be processed. Job queues may store task data measured in megabytes.

2. Joint Dataset

A dataset created by combining multiple sources. Joint datasets often increase total size in megabytes.

3. JSON Data Size

The amount of storage required for data stored in JSON format. JSON files can grow quickly in megabytes.

4. Jitter Buffer

A temporary storage area that smooths data flow. Jitter buffers store megabytes during real-time processing.

5. Job Scheduler

A system that decides when tasks run. Schedulers manage memory and data measured in megabytes.

6. Just-in-Time Processing

Processing data only when needed. This approach can reduce how many megabytes are stored at once.

7. Joint Memory Allocation

Memory shared across multiple AI tasks. Joint allocation must carefully manage available megabytes.

8. Java Bytecode Size

The storage size of compiled Java programs. Bytecode files occupy megabytes on disk.

9. Job Metadata

Information describing AI tasks. Metadata adds additional megabytes to total system storage.

10. Junction Storage Point

A logical link between storage locations. Junctions help organize data measured in megabytes.

Letter K

1. Kernel Memory

Memory reserved for core system operations. Kernel memory usage is tracked in megabytes.

2. Key-Value Store

A data storage method that pairs keys with values. Large key-value stores consume many megabytes.

3. Knowledge Base Size

The total amount of stored knowledge in an AI system. Size is measured in megabytes or larger units.

4. K-Means Dataset Size

The amount of data used in clustering algorithms. Dataset size affects memory usage in megabytes.

5. Keras Model Size

The storage required for AI models built with Keras. Model size is measured in megabytes.

6. Keyframe Data

Important frames selected from video data. Keyframe storage contributes to total megabyte usage.

7. Kernel Cache

A cache used by the operating system kernel. Kernel caches store frequently used data in megabytes.

8. Knowledge Graph Storage

Memory used to store relationships between data points. Knowledge graphs require significant megabytes.

9. Kinetic Data Stream

Continuously changing data from sensors or systems. These streams generate growing megabytes over time.

10. K-Nearest Neighbor Memory Usage

The memory required by nearest-neighbor algorithms. Memory use increases as datasets grow in megabytes.

Letter L

1. **Latency**
   The delay between requesting data and receiving it. Latency affects how quickly megabytes move through AI systems.

2. **Layer Output Size**
   The amount of data produced by a neural network layer. Output size is measured in megabytes during processing.
3. **Load Balancing**
   Distributing work across systems so no single device is overloaded. Balancing helps manage megabytes efficiently.
4. **Local Storage**
   Data stored on a device rather than the cloud. Local storage capacity is measured in megabytes.
5. **Log File Size**
   The amount of storage used by system logs. Logs can accumulate many megabytes over time.
6. **Loss History Data**
   Stored values showing how model error changes during training. Loss history consumes megabytes.
7. **Low-Precision Data**
   Data stored using fewer bits per value. Low-precision formats reduce megabyte usage.
8. **Labeled Dataset**
   Data that includes tags or labels for AI training. Labels increase total dataset size in megabytes.
9. **Linear Model Size**
   The storage required for a linear AI model. Size is measured in megabytes.
10. **Load Time**
    The time required to bring data or models into memory. Load time depends on megabytes involved.

Letter M

1. **Memory Allocation**
   Assigning a specific amount of memory for tasks. Allocation is tracked in megabytes.
2. **Memory Footprint**
   The total memory an AI program uses. Footprint is measured in megabytes.
3. **Metadata Size**
   The storage used by descriptive information about data. Metadata adds extra megabytes.
4. **Mini-Batch Size**
   The amount of data processed at once during training. Batch size affects megabyte usage.
5. **Model Checkpoint Size**
   The storage required for saved training states. Checkpoints often consume many megabytes.
6. **Model Compression**
   Reducing model size while preserving performance. Compression lowers megabytes required.
7. **Model Parameters**
   Values learned by an AI model. Parameter storage contributes to total megabytes.
8. **Memory Bandwidth**
   The rate at which memory transfers data. Bandwidth affects how fast megabytes move.

9. **Multimodal Dataset**
   Data combining text, images, audio, or video. Multimodal data uses large megabytes.
10. **Mutable Data**
   Data that changes during processing. Mutable data can increase temporary megabyte usage.

Letter N

1. **Network Bandwidth**
   The capacity of a network connection. Bandwidth controls how many megabytes can be transferred.
2. **Neural Network Size**
   The total storage required for a neural network. Size is measured in megabytes.
3. **Node Memory**
   Memory available on a computing node. Node memory limits megabytes per task.
4. **Normalization Data**
   Values used to scale data consistently. Normalization adds to dataset megabytes.
5. **Non-Volatile Memory**
   Memory that retains data without power. Capacity is measured in megabytes.
6. **Numerical Precision**
   The detail level of stored numbers. Higher precision increases megabytes.
7. **Noise Data**
   Unwanted or random information in datasets. Noise increases data size in megabytes.
8. **Neural Activation Size**
   Memory used to store neuron outputs. Activation size is measured in megabytes.
9. **Network Transfer Size**
   The amount of data sent across a network. Transfer size is tracked in megabytes.
10. **Node Cache**
   Temporary storage on a node for faster access. Cache size uses megabytes.

Letter O

1. **Object Storage**
   A storage method that manages data as objects. Capacity is measured in megabytes.
2. **Offline Dataset**
   Data stored locally rather than streamed. Offline datasets occupy megabytes.
3. **On-Device Storage**
   Storage located on the AI device itself. On-device capacity is measured in megabytes.
4. **Operational Memory**
   Memory used during active AI operations. Usage is tracked in megabytes.
5. **Optimization Data**
   Information used to improve model performance. Optimization data uses megabytes.

6. **Output Buffer**
   Temporary storage for results before delivery. Buffers hold megabytes.
7. **Overhead Data**
   Extra data required for system operation. Overhead increases total megabytes.
8. **Object Detection Output Size**
   The amount of data produced by detection models. Output size is measured in megabytes.
9. **Online Storage**
   Storage accessible via the internet. Online storage limits are measured in megabytes.
10. **Open Dataset Size**
    The storage footprint of publicly available datasets. Size is measured in megabytes.

Letter P

1. **Parameter Count**
   The total number of model values. Count correlates with megabytes required.
2. **Persistent Storage**
   Long-term storage that retains data. Capacity is measured in megabytes.
3. **Pipeline Buffer**
   Temporary storage between processing stages. Buffers manage megabytes in motion.
4. **Preprocessing Data**
   Data created while preparing datasets. Preprocessing adds megabytes.
5. **Precision Reduction**
   Lowering numeric detail to save space. Reduction decreases megabyte usage.
6. **Prediction Output Size**
   The amount of data produced by model predictions. Size is measured in megabytes.
7. **Primary Memory**
   Main system memory used for processing. Capacity is measured in megabytes.
8. **Processing Overhead**
   Extra resources needed during computation. Overhead increases megabyte usage.
9. **Profiling Data**
   Information collected to analyze performance. Profiling data uses megabytes.
10. **Paged Memory**
    Memory divided into fixed-size blocks. Paging manages megabytes efficiently.

Letter Q

1. **Quantization**
   A technique that reduces numerical precision to lower memory usage. Quantization helps AI models use fewer megabytes.
2. **Query Cache**
   Stored results from previous searches or requests. Query caches save megabytes by avoiding repeated computation.

3. **Query Latency**
   The time it takes to retrieve data after a request. Latency depends on how many megabytes must be accessed.
4. **Queue Buffer**
   A temporary holding area for data waiting to be processed. Buffers store megabytes in order.
5. **Quick Access Memory**
   Memory designed for fast retrieval. It handles megabytes needed immediately by AI tasks.
6. **Quantitative Dataset**
   Data made up of numerical values. These datasets are stored and measured in megabytes.
7. **Query Result Size**
   The amount of data returned from a query. Result size is measured in megabytes.
8. **Quorum Storage**
   A distributed storage method requiring agreement across systems. Data is replicated in megabytes.
9. **Quality Metrics Data**
   Information used to measure model performance. Metrics consume storage in megabytes.
10. **Quantized Model Size**
    The reduced size of an AI model after quantization. Size is typically measured in megabytes.

Letter R

1. **RAM Usage**
   The amount of active memory in use. RAM usage is tracked in megabytes.
2. **Read Throughput**
   The speed at which data is read from storage. Throughput affects how fast megabytes are accessed.
3. **Real-Time Processing**
   Processing data immediately as it arrives. Real-time systems manage megabytes continuously.
4. **Recursive Memory Use**
   Memory consumption during repeated function calls. Usage accumulates in megabytes.
5. **Redundant Storage**
   Extra copies of data kept for reliability. Redundancy increases total megabytes stored.
6. **Resource Allocation**
   Assigning system resources like memory and storage. Allocation is measured in megabytes.
7. **Runtime Memory**
   Memory required while a program is running. Runtime memory usage is tracked in megabytes.
8. **Raw Dataset Size**
   The size of unprocessed data. Raw datasets often require many megabytes.

9. **Read-Only Memory Footprint**
   The storage space used by non-modifiable data. Footprint is measured in megabytes.
10. **Retrieval Speed**
   How quickly stored data can be accessed. Speed depends on the number of megabytes involved.

Letter S

1. **Sample Size**
   The number of data examples used in AI training. Larger samples increase megabytes.
2. **Scalable Storage**
   Storage that grows with demand. Scalability supports increasing megabytes.
3. **Serialized Data Size**
   The storage size of data converted into a transferable format. Size is measured in megabytes.
4. **Shared Memory**
   Memory accessible by multiple processes. Shared memory usage is tracked in megabytes.
5. **Snapshot Size**
   The storage required for saved system states. Snapshots consume megabytes.
6. **Sparse Data Representation**
   A storage method that saves space when data contains many empty values. Sparsity reduces megabytes.
7. **Storage Capacity**
   The total amount of data a system can store. Capacity is measured in megabytes or more.
8. **Streaming Data Buffer**
   Temporary storage for incoming data streams. Buffers handle megabytes in motion.
9. **System Cache**
   Memory that stores frequently used data. Caches manage megabytes for faster access.
10. **Synchronization Data**
   Information used to keep systems aligned. Sync data adds to total megabytes.

Letter T

1. **Temporary Storage**
   Short-term storage used during processing. Temporary storage holds megabytes briefly.
2. **Tensor Size**
   The memory required to store multidimensional data structures. Tensor size is measured in megabytes.
3. **Throughput Rate**
   The amount of data processed per unit time. Rate is often described in megabytes per second.

4. **Training Dataset Size**
   The total amount of data used to train AI models. Size is measured in megabytes.
5. **Transfer Buffer**
   Temporary memory used during data movement. Buffers hold megabytes during transfer.
6. **Transactional Storage**
   Storage that ensures data consistency. Transactions add overhead measured in megabytes.
7. **Thread Memory Usage**
   Memory consumed by execution threads. Usage is tracked in megabytes.
8. **Token Storage Size**
   The memory required to store processed tokens. Token data uses megabytes.
9. **Time-Series Data Size**
   The storage footprint of data collected over time. Size increases in megabytes.
10. **Training Checkpoint Size**
    The storage required for saved training states. Checkpoints are measured in megabytes.

Letter U

1. **Unified Memory**
   A memory system shared between CPU and GPU. Unified memory manages data in megabytes efficiently across processors.
2. **Upload Size**
   The amount of data sent to storage or the cloud. Upload size is measured in megabytes.
3. **User Data Storage**
   Storage allocated for user-generated files. Capacity is tracked in megabytes.
4. **Utilization Rate**
   The percentage of memory or storage currently in use. Utilization reflects how many megabytes are occupied.
5. **Uncompressed Data Size**
   The size of data before compression. Uncompressed data often uses more megabytes.
6. **Update Payload**
   Data sent during software or model updates. Payload size is measured in megabytes.
7. **Uptime Storage Logs**
   Logs recorded while systems run continuously. Logs accumulate megabytes over time.
8. **Unified Dataset**
   A dataset combined from multiple sources. Size is measured in megabytes.
9. **User Session Memory**
   Memory allocated for active users. Session memory uses megabytes dynamically.
10. **Usage Quota**
    A storage limit set for users or systems. Quotas are defined in megabytes or larger units.

Letter V

1. **Validation Dataset Size**
   The amount of data used to evaluate AI models. Size is measured in megabytes.

2. **Vector Embedding Size**
   The memory required to store numerical representations of data. Size is measured in megabytes.
3. **Virtual Memory**
   A system that extends RAM using disk storage. Virtual memory is managed in megabytes.
4. **Volatile Memory**
   Memory that loses data when power is off. Capacity is measured in megabytes.
5. **Video Dataset Size**
   The storage required for video data. Video datasets consume many megabytes.
6. **Versioned Storage**
   Storage that keeps multiple file versions. Versioning increases megabyte usage.
7. **Vision Model Size**
   The memory footprint of computer vision models. Size is measured in megabytes.
8. **Vector Database Storage**
   Storage used for similarity search data. Vector databases consume large megabytes.
9. **Virtualized Storage**
   Storage abstracted from physical hardware. Capacity is tracked in megabytes.
10. **Volume Capacity**
    The total size of a storage volume. Capacity is measured in megabytes or more.

Letter W

1. **Weight Parameters**
   Learned values in AI models. Weight storage contributes to megabytes used.
2. **Working Memory**
   Memory actively used during processing. Working memory usage is measured in megabytes.
3. **Write Throughput**
   The speed at which data is written to storage. Throughput affects megabytes per second.
4. **Warm Cache**
   Cached data that is already loaded. Warm caches reduce repeated megabyte transfers.
5. **Workflow Data Size**
   The total data used across an AI workflow. Size is measured in megabytes.
6. **Web Dataset Size**
   The storage required for web-based data. Size is tracked in megabytes.
7. **Weight Compression**
   Reducing model weight size to save memory. Compression lowers megabytes used.
8. **Window Size (AI Context)**
   The amount of data processed at once. Window size affects memory megabytes.
9. **Writable Storage**
   Storage that allows data modification. Capacity is measured in megabytes.
10. **Workload Memory Demand**
    The memory required for tasks. Demand is expressed in megabytes.

Letter X

1. **X-Axis Data Size**
   Data stored along one dimension of a dataset. Size contributes to megabytes used.
2. **XML Data Size**
   The storage footprint of XML-formatted data. XML files often use many megabytes.
3. **XOR Encoding**
   A binary operation used in data processing. Encoded data occupies megabytes.
4. **Execution Context Size**
   Memory required during task execution. Size is measured in megabytes.
5. **External Dataset Size**
   The size of data sourced externally. External datasets add megabytes.
6. **Expanded Memory Use**
   Memory growth during processing. Expansion increases megabytes consumed.
7. **Experimental Dataset**
   Data used for testing AI models. Size is measured in megabytes.
8. **Extraction Buffer**
   Temporary storage during data extraction. Buffers store megabytes briefly.
9. **Extended Storage Capacity**
   Additional storage added to a system. Capacity is measured in megabytes.
10. **Explainability Data Size**
    Data used to explain AI decisions. Explainability outputs consume megabytes.

Letter Y

1. **Yield Data Size**
   The amount of useful output data produced. Yield is measured in megabytes.
2. **Yearly Data Growth**
   Annual increase in stored data. Growth is tracked in megabytes.
3. **YAML File Size**
   The storage footprint of YAML configuration files. Size is measured in megabytes.
4. **Y-Axis Feature Size**
   Feature data along a dataset dimension. Size contributes to megabytes.
5. **Young Generation Memory**
   Memory area for newly created data. Usage is measured in megabytes.
6. **Yield Optimization Data**
   Data used to improve output efficiency. Stored in megabytes.
7. **YourAIStudyBuddy Dataset Size**
   The total data used by AI study tools. Size is measured in megabytes.
8. **Yottabyte Reference Scale**
   A conceptual upper scale of data measurement. Used for comparison beyond megabytes.
9. **Yield Metrics Storage**
   Storage used for performance metrics. Metrics consume megabytes.
10. **Yoked Memory Allocation**
    Linked memory regions used together. Allocation is measured in megabytes.

Letter Z

1. **Zero-Copy Memory**
   A method that avoids duplicating data. Zero-copy reduces megabytes in motion.
2. **Zipped Dataset Size**
   The storage size after compression. Zipped files use fewer megabytes.
3. **Zettabyte Scale (Reference)**
   A large-scale data concept. Used to compare growth beyond megabytes.
4. **Zone-Based Storage**
   Storage divided into regions. Zones manage megabytes efficiently.
5. **Z-Score Normalization Data**
   Statistical values used in preprocessing. Data adds megabytes.
6. **Zero Padding**
   Adding placeholder data to inputs. Padding increases memory megabytes.
7. **Zoomed Feature Maps**
   Scaled representations in vision models. Stored in megabytes.
8. **Zonal Cache**
   Cache divided by region. Cache size is measured in megabytes.
9. **Zero-Day Model Update Size**
   The storage required for emergency updates. Size is measured in megabytes.
10. **Zipped Model Archive**
    Compressed AI model files. Archives reduce megabytes required.

Appendix A

Understanding AI Infrastructure: The Foundation Behind Intelligent Systems

Physical Infrastructure

Component	Purpose
Data Centers	House AI computing systems
Servers	Run AI workloads
GPUs	Perform AI calculations
CPUs	Manage system operations
Storage Systems	Store data and models
Fiber Optic Networks	Move information

Component	Purpose
Cooling Systems	Prevent overheating
Power Systems	Supply electricity
Semiconductor Factories	Manufacture AI chips
User Devices	Access AI services

Digital Infrastructure

Component	Purpose
Operating Systems	Manage hardware resources
AI Models	Perform intelligent tasks
Algorithms	Guide decision-making
Databases	Store information
Cloud Platforms	Deliver AI services
APIs	Connect systems
Training Datasets	Teach AI systems
Machine Learning Frameworks	Build AI models
Cybersecurity Systems	Protect digital assets
User Interfaces	Allow human interaction

Big Idea

Artificial intelligence is not just software. AI depends on a vast network of physical infrastructure and digital infrastructure working together. Without data centers, chips, networks, storage systems, software, algorithms, and electricity, modern AI systems would not exist.

Teacher Implementation Guide

Large Language Models: How Machines Learn Human Language

Purpose of This Teacher Guide

This Teacher Implementation Guide is designed to support educators who are introducing students to the foundational concepts of **Large Language Models (LLMs)** and artificial intelligence language systems.

While the student book explains the science and ideas behind large language models, this guide provides educators with practical strategies to help students:

- Understand how machines learn language
- Develop critical thinking about artificial intelligence

• Explore the benefits and limitations of AI systems
• Reflect on the ethical and societal impacts of intelligent technologies

The goal of this course is not simply to teach students how to use AI tools, but to help them **understand how these systems work and how they should be used responsibly**.

Large language models represent one of the most significant technological developments of the modern era. By studying these systems, students gain insight into how knowledge, communication, and learning may evolve in the future.

Recommended Grade Levels

This book is designed for:

• **Grades 6–12**
• **Early college or introductory technology courses**
• **AI literacy programs and STEM enrichment programs**

The material is written to be accessible to students with **no prior programming experience**.

Course Length Options

Teachers may implement this curriculum in several formats depending on available instructional time.

Option 1 — Full Semester Course

16 weeks
1 chapter per week

Option 2 — Quarter Course

8 weeks
2 chapters per week

Option 3 — AI Literacy Module

4 weeks
Selected chapters

Suggested Weekly Teaching Model

Each chapter can be taught using a consistent instructional structure:

1. Opening Discussion (10 minutes)

Introduce the central idea of the chapter with a question.

Examples:

- How do humans learn language?
- Can machines truly understand words?
- What makes communication meaningful?

This helps activate prior knowledge and curiosity.

2. Concept Exploration (20–30 minutes)

Students read or review the chapter content.

Teachers may:

- Read sections aloud
- Assign small reading groups
- Highlight key concepts on the board

Encourage students to identify:

- Key terms
- Important concepts
- Real-world applications

3. Classroom Discussion (15 minutes)

Discuss major ideas from the chapter.

Possible prompts:

- Why is language difficult for machines to understand?
- How might AI language systems change education?
- What risks could exist if AI produces incorrect information?

4. Activity or Worksheet (15–20 minutes)

Students complete a worksheet or activity designed to reinforce understanding.

Examples:

- Tokenization exercises
- Pattern recognition activities

• AI ethics debates
• Technology case studies

5. Reflection (10 minutes)

Students reflect on what they learned.

Reflection questions may include:

• What surprised you most about this chapter?
• What new questions do you have about AI?
• How might this technology affect your future?

Learning Goals of the Book

By the end of this course, students should be able to:

Explain how human language can be transformed into digital data.

Describe how machines analyze language using tokens and patterns.

Understand the role of neural networks and transformers in AI systems.

Explain how parameters allow large language models to store knowledge.

Evaluate the benefits and limitations of AI language systems.

Analyze ethical challenges related to artificial intelligence.

Discuss how AI may shape the future of work, education, and communication.

Chapter Implementation Overview

Stage 1 — Understanding Language

Chapters 1–4

Focus: Foundations of human language and how it becomes machine-readable data.

Teacher Goals:

• Help students understand why language is complex
• Introduce the concept of digital representation of language
• Explain tokens and pattern recognition

Suggested Activities:

• Sentence tokenization exercise
• Grammar pattern recognition activity
• Discussion on how humans interpret meaning

Key Questions:

• What makes language difficult for machines to understand?
• How do humans interpret context differently than computers?

Stage 2 — How Machines Learn Language

Chapters 5–9

Focus: The technological systems behind language models.

Teacher Goals:

• Introduce neural networks and machine learning
• Explain transformer architecture
• Help students understand parameters and training data

Suggested Activities:

• Visual diagram of neural networks
• Pattern prediction exercises
• AI training simulation activity

Key Questions:

• How do machines learn patterns from data?
• Why do large language models require enormous datasets?

Stage 3 — How AI Thinks With Language

Chapters 10–12

Focus: Context understanding, reasoning, and limitations.

Teacher Goals:

• Teach students how AI predicts language
• Explain hallucinations and model errors
• Encourage critical thinking about AI outputs

Suggested Activities:

• Compare human vs AI responses
• Analyze examples of AI hallucinations
• Evaluate reliability of AI-generated information

Key Questions:

• Why do language models sometimes produce incorrect answers?
• Should people always trust AI responses?

Stage 4 — Society, Ethics, and the Future

Chapters 13–16

Focus: Responsible AI use and future implications.

Teacher Goals:

• Encourage ethical thinking about AI systems
• Discuss social impacts of intelligent technology
• Prepare students to use AI responsibly

Suggested Activities:

• AI ethics debate
• Group discussion on AI in education
• Future technology prediction project

Key Questions:

• Who should control AI systems?
• How can societies ensure AI benefits humanity?

Assessment Suggestions

Teachers may evaluate student learning through several methods.

Formative Assessments

- Chapter quizzes
- Reflection responses
- Group discussions
- Worksheets

Summative Assessments

- Research presentations
- Essays on AI ethics
- Final project: Designing a responsible AI system

Classroom Discussion Topics

Teachers may encourage critical thinking using questions such as:

- Should AI systems be allowed to generate school assignments?
- How might AI change jobs in the future?
- What responsibilities do developers have when creating AI systems?

Final Project Idea

Students may complete a project titled:

"Designing an AI System for the Future."

Students describe:

- What their AI system would do
- What problem it solves
- How it would protect users from harm
- What ethical rules should guide it

This helps students apply both **technical knowledge and ethical reasoning**.

Encouraging Responsible AI Literacy

One of the most important goals of this book is to help students become **informed and responsible users of artificial intelligence**.

Students should learn that:

- AI systems are powerful tools but not perfect sources of truth
- Human judgment is essential when using AI
- Ethical decision-making must guide technological development

By understanding how large language models work, students gain the knowledge needed to navigate a future where humans and intelligent machines collaborate more closely than ever before.

Final Note to Educators

Artificial intelligence is transforming how knowledge is created, shared, and understood. Teaching students about large language models is not only about technology—it is about preparing the next generation to think critically about the tools shaping their world.

This book provides a foundation for that understanding.

With thoughtful guidance, educators can help students move beyond simply using AI tools toward **understanding the science, responsibilities, and possibilities behind them**.

NOTES:

Notes:

www.ingramcontent.com/pod-product-compliance
Lightning Source LLC
Chambersburg PA
CBHW041754220926
51847CB00014B/200